The Battle for Wardenclyffe

a story in letters

By Ernst Willem van den Bergh

Contents

Introduction

Some historical events may appear different from the outside than they were experienced at that time by those directly involved. The battle for Wardenclyffe certainly is one such event.

In one corner we have Nikola Tesla, often portrayed as "the Mad Scientist" who wanted to build a huge tower that would become the first part of his envisioned "World System". This system would provide a worldwide communication network, just like the internet today, and a navigation system, think of GPS today. But more than that, it would also provide worldwide wireless power and if you read his work carefully you will find that it would also *generate* power from atmospheric electricity induced by cosmic rays. Tesla's Magnifying Transmitter would provide a system so complete that it would advance humanity at least a century.

In the other corner we find John Pierpont Morgan, one of the wealthiest men of his time, who was interested in worldwide communication, but did not care much for all that other stuff. In fact, that other stuff may interfere with his other businesses in a way he could not predict.

When Tesla returned from Colorado Springs where he had found proof that his plan would work, he went to Morgan for the money this required. Unfortunately Tesla hugely underestimated the actual costs and Morgan's actions on the markets caused the prices to rise, and then there was Guglielmo Marconi who accomplished the first wireless transatlantic communication, reducing Morgan's interest in Tesla's system to zero.

With this book I want to show the world what, up until today, only a select few had access to; the story told by the letters exchanged between Tesla and Morgan. They tell the story as it was felt at that time.

This has been made possible with the help of many contributions at www.energeticforum.com to my project to transcribe many of the letters of Nikola Tesla that were saved on microfilm.

Not all letters could be read as can be deduced from their numbering, but I think there is enough here to paint a clear picture.

As a token of gratitude I will print the real name or forum-name of the one who did the transcription with every letter.

Setting the stage

Tesla wanted to provide power to the entire world. He figured there were 3 ways to accomplish that.

- Burning some sort of fuel
- Use the energy of the sun stored in the ambient medium
- To transmit, through the medium, the sun's energy to distant places from some locality where it was obtainable

The first option he considered to be not a real solution as it uses up limited resources and produces waste.

This leaves us with only two real options.

Initially Tesla believed that transmitting industrial amounts of energy was not really possible so he focussed on the second option.

Lightning proves that there is plenty of energy in the air if only we could gain access to it *before* it turns into lightning.

This is why Tesla seemed so obsessed with generating extremely high voltages; he wanted to mimic lightning so to tap its power.

While in Colorado Springs doing his research he discovered something that changed his mind and consequently his plans.

We read in Tesla's own words in his article of March 5th, 1904 *"The Transmission of Electrical Energy Without Wires"* published in the *"Electrical World and Engineer"*:

"It was on the third of July - the date I shall never forget - when I obtained the first decisive experimental evidence of a truth of overwhelming importance for the advancement of humanity. A dense mass of strongly charged clouds gathered in the west and towards the evening a violent storm broke loose which, after spending much of its fury in the mountains, was driven away with great velocity over the plains. Heavy and long persisting arcs formed almost in regular time intervals.

My observations were now greatly facilitated and rendered more accurate by the experiences already gained. I was able to handle my instruments quickly and I was prepared. The recording apparatus being properly adjusted, its indications became fainter and fainter with the increasing distance of the storm, until they ceased altogether. I was watching in eager expectation. Surely enough, in a little while the indications again began, grew stronger and stronger and, after passing through a maximum, gradually decreased and ceased once more. Many times, in regularly recurring intervals, the same actions were repeated until the storm which, as evident from simple computations, was moving with nearly constant speed, had retreated to a distance of about three hundred kilometres. Nor did these strange actions stop then, but continued to manifest themselves with undiminished force. Subsequently, similar observations were also made by my assistant, Mr. Fritz Lowenstein, and shortly afterwards several admirable opportunities presented themselves which brought out, still more forcibly, and unmistakably, the true nature of the wonderful phenomenon. No doubt, whatever remained: I was observing stationary waves.

As the source of disturbances moved away the receiving circuit came successively upon their nodes and loops. Impossible as it seemed, this planet, despite its vast extent, behaved like a conductor of limited dimensions. The tremendous significance of this fact in the transmission of energy by my system had already become quite clear to me. Not only was it practicable to send telegraphic messages to any distance without wires, as I recognized long ago, but also to impress upon the entire globe the faint modulations of the human voice, far more still, to transmit power, in unlimited amounts, to any terrestrial distance and almost without loss."

As the plan to derive power from the atmosphere was already far advanced, now Tesla could add a wireless distribution system to it. And so Tesla finally settled for the third option by building a network of towers that housed his "Magnifying Transmitter" and that could send and power to small receivers all around the world. He knew that financiers would either doubt this possibility or fear it. But many were interested in worldwide wireless communication and if you can send power it is but a small step to modulate that power and thus send messages.

With this plan Tesla thought he could easily convince Morgan and once he would see all the other things Tesla's system could do, he would surely applaud.

The Magnifying Transmitter

A few words on Tesla's Magnifying Transmitter are in order here.

The general consensus today is that this is simply a Tesla coil with a so called "extra coil" added, as is described in patent 1,119,732 "Apparatus for Transmitting Electrical Energy".

Other than the fact that the picture in this patent vaguely resembles the Wardenclyffe tower, there is absolutely no reason to assume that this is true.

In fact, if you read everything that Tesla wrote about his Magnifying Transmitter you will find that it doesn't match *at all* with this patent.

The apparatus in this patent transmits energy, but it does not magnify it. It only transmits what you put into it.

The Magnifying Transmitter magnifies the input energy and *then* transmits it.

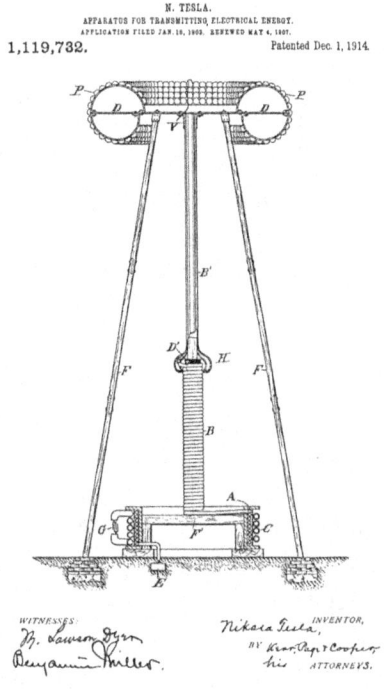

Leland Anderson provided *"Rare notes on Wardenclyffe"* to the *"Electric Spacecraft"* which they printed in issue 26 in 1997. These notes show the actual diagram of a Magnifying Transmitter. One of these has also been printed in *"From Colorado Springs to Long Island"* published by the Tesla Museum in Belgrade, along with a number of variations.

Also the *"Colorado Springs Notes"* clearly show a number of variations on this theme.

For fear of copyright infringements I have not copied these diagrams here. Instead I will show the difference using my simplified diagram below.

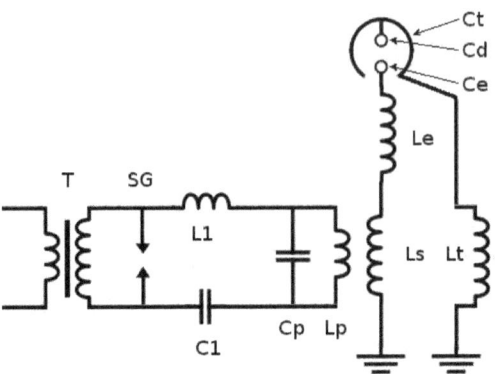

It starts with a primary high voltage source (T), a spark-gap (SG), a primary capacitor (C_p) and a primary coil (L_p), so far it is the circuit of a usual Tesla coil. Then there is a secondary coil (L_s) and an *optional* so called "extra coil" (L_e).
Up to here it more or less matches with the before mentioned patent, but here is the difference:
A second spark-gap between C_e and C_d, C_d then connects to the "*cupola*", the roof of the tower which is in turn connected through an inductor (L_t) to the ground.

It would go too far to dive into the details here. I will do so in another publication, but it is easy to see that there are a number of very significant differences.

Let's now turn to the conversation between Nikola Tesla and John Pierpont Morgan.

Disclaimer from the transcription team: Some of the letters, telegrams or handwriting are hardly legible; too dark or washed out. We pride ourselves to be as accurate as possible in guessing words that are not really legible. Words or parts thereof that we could not guess have been replaced with dots (…).

So it began...

JPM 0001 — transcript by Jeff Pearson

New York, Nov. 26th, 1900.
46 & 48 East Houston Str.

Mr. J. Pierpont Morgan,
        Wall Street, New York:

Dear Mr. Morgan:-
        Since last Friday, when I took the liberty to bring my
project to your kind attention, I have carefully considered the
facts and features, which seem to me essential and which, briefly
stated, are as follows:
        1. I have succeeded in perfecting methods and apparatus,
which permit the transmission of messages to any distance without
wires ....long and expansive cables, as a method of conveying
intelligence, a commercial impossibility. These inventions render
....practicable, the production and safe manipulation of electrical pressures up to a hundred million volts and movements of electrical energy measured by rates of hundred of thousands of horse-
power which are capable of bringing into action instruments at any
point of the globe, no matter how distant from the transmitting
station. Long practical experience with apparatus of this kind and
exact measurements embracing a range of nearly seven hundred miles,
.... me to construct plants for such telegraphic communication
.... Atlantic and, if need be, across the pacific ocean, with
the fullest assurance of success.
        2. I have devised means for operating selectively a
great number of instruments without mutual interference, and can
guarantee the absolute privacy of all messages.
        3. rights have been secured by me of these methods
.... patents granted or allowed.
        4. I am free and unhampered to enter into such agree-
ments regarding the exploitation of these, and my future inventions
in this field, as you might desire.
        5. It would be my wish, subject to your approval, to
identity my name with any corporation which might be eventually
formed for above purpose.
        6. Although the development of these inventions has con-
sumed .... of effort, knowing that I have to deal with a great
man, I do not hesitate to leave the apportionment of my interest
and compensation .... to your generosity.

JPM 0002 — transcript by Jeff Pearson

7. The first necessity is the provision of a fund to cover the expenses of construction of two plants for trans-atlantic or trans-pacific communication.

8. With temporary buildings the former would involve an expenditure of about one hundred thousand dollars, the latter approximately two hundred and fifty thousand.

9. The working capacity of the atlantic plants would be equal to at least that of four of the present cables. The pacific plants would do the work of more than eight trans-pacific cables

10. Six to eight months would be necessary for the creation of the former, and the latter could be put in working order certainly within a year.

Hoping in the interest of the art, if not in my own, that you will ..... consider this project, the realisations which ....quite busy by my inventions, I have the honor to be

Yours very respectfully

Tesla proposes his plan to Morgan. Two towers, one for transatlantic and one for transpacific communication. For a total of 350,000 dollars. I do not understand why Tesla put it this way. With the discoveries that he claimed, one single tower could cover the entire world. The only reason *that I see* for additional towers at the same location would be that he needed different frequencies for his 'art of individualization'.

Tesla was invited to elaborate on Friday, Dec 7th, 1900.

New York, Dec. 10Th, 1900.
46 & 48 East Houston St.

Mr. J Pierpont Morgan,
     23 Wall Street,
        New York City

Dear Mr. Morgan:-
     Appreciating the immense value of your time and also
the weight of circumstantial ...., I have withdrawn more or less
hastily last Friday, preferring to make a few condensed statements
at long distance which, with a small effort on your part will put
you in possession of that knowledge; which I have myself gained
only after a long and exhaustive study of the questions involved
     1. A casual remark ... made the other day showing that
you have been somewhat impressed by newspaper reports, skillfully
..... , but without foundations in facts ....
..... , Priv. Councillor of Germany: "I am devoting myself
...... sometime to investigations in wireless telegraphy which you
have first founded in such a clear and precise manner"--- "It will
interest you, as father of this telegraphy, to know," ---
     2. Apologizing for this digression, which I would have
..... had I not thought it necessary to illuminate the situation.
I beg you to bear in mind, that with my patents in this still vir-
gin field, should you take hold of them, you will command a position which, for a number of reasons, will be ...... stronger than
that held by the owners of Bell's telephone inventions or by those
of my own discoveries in power transmission by alternating currents.
     3. For not only have I secured broad rights on the fun-
damental elements and features of the system, but I also have patents ...., methods .... appliances for the
production of powerful oscillations, which are universally identified with my name, and without which a wireless transmission to
very great distances and individualization of messages are not
practicable. The necessity of ...... my statements compels
me to cite my illustrious friends, Lord Kelvin and Sir William
Crookes. The former says: "This is a wonderful development of the
induction coil and destined to be of great importance!." the latter: "The performance of your machine is marvellous."
     4. The legal position is rendered still more secure by
my inventions of entirely novel methods for receiving as well as
for individualizing the messages, with which none now known has the
remotest chance of competing, and which are likewise patented.

JPM 0004 — transcript by Jeff Pearson

5. Furthermore, I have secured rights on improvements of a wholly different nature applicable to my system and offering such advantages that any of them would be sufficient to exclude competition.

6. Then again, I have made certain scientific discoveries, which in themselves are of transcendent importance in wireless transmissions to great distances, which I chiefly contemplate, and which offers such unprecedented changes in commercial exploitation, that they deserve your fullest attention

7. Without endeavoring to give you more evidences of this kind, I venture to state -not without apprehension of being misunderstood to you- that you have in further guarantee of success in my extensive experience and ability acquired through years of experimentation with electrical vibrations, of which my co-workers will give you testimony.

8. Besides in this country, I have protected myself, though not so quite completely, in England, Victoria, New South Wales, Austria, Hungary, Germany, France, Italy, Belgium, Russia and Switzerland.

9. In view of the intense activity in this field it is desirable, that I should be placed without delay in a position enabling me to turn to profit my advanced knowledge and to secure patents on features, which though of less moment now, may soon become important.

10. Before going further permit me to remind you, that had there been only faint-hearted and close-fisted people in the world, nothing great would have ever been done. Rafael could not have created his marvels, Columbus could not have discovered America, the Atlantic cable could not have been laid. You of all, should be the man to embark boldly in this enterprise only seemingly hazardous, prompted by superior insight as well as desire to advance an art of inestimable value to mankind.

11. Coming to the financial question - please remember: These inventions, - the results attainable only by their means - which now I alone am able to accomplish - in your strong hands, with your consummate knowledge the mastery of business - are worth an incalculable amount of money.

12. Although I have expressed myself in my last letter, I will be more explicit relative to my share and compensation, The control is yours, the larger part is yours. As to my interest - you know the value of discoveries and artist creations - your terms are mine.

13. Referring to your remark about the state of affairs in England, the acceptance by the Post office of the Hertzian system which, by the way, is of no practical value, since it does not permit transmission to great distances and selective signalling, would, if anything, help my inventions. But if the government should adapt certain modifications, which are infringements on my system, I would be again in a good position. I have, however, even .... from England, France, or Belgium, that I shall have no difficulty whatever to place and operate my apparatus.

Yours very respectfully,

Apparently Morgan responded with a question about others that were working on wireless transatlantic communication. To which Tesla responds with my way is the only way. Which actually turned out to be true when the US Supreme Court ruled in 1943 that all Marconi's radio patents were invalid as everything was described in Tesla's patents pre-dating Marconi's.

Morgan agrees to fund Tesla's plan.

JPM 0006 — transcript by Jeff Pearson

Dec 12 1900

How can I begin to thank you
in the name of my profession and
my own great generous man. My work
will proclaim loudly your name to the world!
  You will soon see that not only am
I capable of appreciating deeply the
nobility of your action, but also of
making your primarily philantrophic
investment worth a hundred times the
sum you have put at my disposal
in such a magnanimous princely way !
With many, many wishes from all my heart
for your happiness and welfare believe me
ever yours Most Faithfully

N. Tesla

Mr. J. P. Morgan
City
Copy of letter
Forwarded

(Handwritten letter)

Finalizing the deal

JPM 0007 — transcript by Jeff Pearson

S T A T E M E N T

Relative to the cash expended by Mr. Tesla in developing the inventions to be introduced up to Jan. 1, 1901.

        According to bills on file:
To general expenses, cost of patents and construction
    of model machines to be manufactured ------------$ 148,466.18

To wages of assistants ----------------------------$ 123,604.05

To work in Colorado -------------------------------$ 16,000.0

To interest acquired April 25, 1898 ---------------$ 158,500.0

To interest acquired Nov. 1, 1900 -----------------$ 56,029.27

        Total amount of cash expended by Mr. Tesla $ 402,599.50

Morgan asked Tesla for an indication of how much he used on his experimental station in Colorado Springs.

JPM 0008 — transcript by Jeff Pearson

New York, February 15th. 1901

Nikola Tesla, Esq.

    46 East Houston St, New York.

My dear Mr.Tesla:-

       I enclose draft of a letter which I think covers the arrangement you talked of with Mr.Morgan. If you will write him such a letter as this and send him at the same time assignments of the fifty-one per cent. interest in the various patents, he can confirm the understanding by letter and that will complete the arrangement. I have shown this to Mr. Morgan and he thinks it is allright.

                  Yours very truly,

                     Charles Steele

Morgan would have agreed on a 50-50 deal, but Tesla offered him 51% hoping that Morgan's business insights will help him forward his plan. Morgan of course accepts but also includes Tesla's patent-rights including the non-related lighting patents all of which were not part of the proposal.

JPM 0009 — transcript by Jeff Pearson

New York, Feb. 18Th, 1901.
46 & 48 East Houston Str.

..Charles Steele
....   J.P. Morgan & co.,
        23 Wall Street, N.Y. City

Dear Mr. Steele:-
        Upon receipt of your kind note of Feb 15th, with formal
letter to Mr. Morgan, confirmatory of our verbal understanding
called on you last Saturday, but learned that you were engaged.
        I need scarcely say that I would sign any document ap-
proved by Mr. Morgan, but I believe there exist a misunderstanding in regard to my system of lighting, which was not included
in the original proposition.
        I have been engaged in perfecting it since a number of
years and believe, that it will create an industrial revolution
The inclosed excerpt from a ..... program which I have .....-
... give you .... of this invention, while the appended
statement relative to the cash expenditures I have incurred, ...
... ... will ... ..., with which you are
no doubt familiar, that takes among other things, capital to develop
.... inventions. Of this ... while it ... been ... for ...-
..., I am proud knowing that the ..... be grateful to ....
it. I am well aware of its great commercial value, but any sum,
however large, which I might obtain from other people, I would consider scarcely more than a ..., as compared with the honor and
satisfaction which I would feel in being associated with Mr. Morgan ..... as soon as practicable, I shall show the light to him
..... take it on his own ......
        Under inclosure please ... my letter, in all respects
..... as you have suggested with the exception of the light
... which, I hope, will be found just and agreeable to him.
        Some of the patents concerned in our agreement are owned
by a company, in which, besides myself, Col. Antor is interested
and which I control. It will be necessary for me to comply with a
formality before making the assignments. I shall attend to this
matter at the earliest possible moment.
        With assurance of my personal sympathy, which has been
kindled in our short conversation the other day, I remain,

        Yours Sincerely

Incls. (Kindly return at your convenience.)

Tesla did not like Morgan's addition of these patents but accepts nevertheless.
Having the Magnifying Transmitter build was the #1 priority.

JPM 0010 — transcript by Jeff Pearson

"TESLA'S ARTIFICIAL DAYLIGHT".

In order to convey a idea of the great commercial possibilities of this invention, attention may be called to a few facts, of which the general public has little knowledge. A common impression is that the electric lighting systems, as now commercially introduced, are a vast advance over older methods of illumination, leaving little room for improvement. This is far from being so, as usually but a fraction of one per cent. of the total energy consumed is obtained in the form of light, all the rest being dissipated in invisible rays, of no value for the purpose. When considering that hundreds of millions of dollars are invested in the United States alone for producing light in these crude ways, the importance of a radical advance in this field will be appreciated.

The present systems, however, are not only wasteful, they are also inseparable from other disadvantages as, the large cost of installation, which is chiefly due to the great quantity of copper required, the frequent renewal of the lamps, owing to their unavoidable deterioration, the disagreeable character of the light which, coming from a small surface, is naturally too intense and detrimental to the eyesight, the necessity of employing more or less opaque screens, which involve a considerable loss in illuminating power, and many other drawbacks of this nature. It is true, that recently Nernst and others have secured some gain in the efficiency of the incandescent lamps or burners by the use of coatings of rare oxides, which permit higher degrees of incandescence. But this departure has not done away with the objectionable features above mentioned, on the contrary, it has only added to them.

In Tesla's new lighting system all these disadvantages are successfully removed. The light is produced with a much smaller expenditure of energy, scarcely more than one eighth of what is presently consumed for the same quantity of light, it is soft and agreeable to the eye, closely resembling daylight, and the cost of the copper conductors is much less. Still better re-sults as regards economy are, however, realizable with this kind of light and are sure to be reached soon in the course of its in-troduction. The lamps are very simple and ornamental and can be manufactured cheaply, need never be renewed and, what is very im-portant, it has been ascertained in long experience that their rays have, like those of the sun, certain germicidal and sanitary properties, so that their use in dwellings will become imperative.

Referring particularly to street-lighting and illumina-tion of public places, Tesla's system offers features of an ideal character. The lamps may be of any desired candlepower and may be adapted to any kind of current of supply, and they last indefinite-ly. The plant is cheaper and more economical and, owing to the softness and diffusive character of the light, the illumination is incomparably better. In view of this it is highly probable, that the arc-light will be driven out of the market entirely, being unable to compete.

- - o O o - -

JPM 0012 — transcript by Jeff Pearson

New York February 25th, 1901

Nikola Tesla, Esq.,

    48 East Houston Street, New York.

My dear Mr.Tesla:-

      I have showed Mr. Morgan your letter of the 18th instant, and he would like to see you on the subject before reaching a conclusion. Would you kindly drop in at your convenience.

                Yours very truly,

                  Charles Steele

Tesla was probably given a chance to exclude his patents here, but he did not. Why would he? His Magnifying Transmitter would give him more money than he would ever need. If Morgan wanted these patents, let him have them!

JPM 0013 — transcript by Jeff Pearson

New York, Feb. 25Th, 1901.
46 & 48 East Houston Str.

Mr. Charles Steele,

    Messrs. J. P. Morgan & Co.,

        23 Wall Street, City.

My dear Mr. Steele:-

    I have just received your kind letter and accordingly will call to-morrow, hoping to find Mr. Morgan disengaged.

        Yours sincerely,

24

New York, March 1rst, 1901.
46 & 48 East Houston Str.

J.P. Morgan, Esg.,
            23 Wall Street
                  New York City

Dear Sir:-
            For several years past I have been engaged in perfecting
a system of electric lighting and in carrying on investigations
relative to wireless telegraphy and telephony, and I have made certain discoveries and inventions in connection with these subjects,
for which patents have been issued to me in the United States and
also in foreign countries.
            I am anxious now to construct the apparatus necessary
for putting my discoveries and inventions to practical use and for
continuing my investigations on the subject named.
            For this I desire to procure the sum of ONE HUNDRED AND
FIFTY THOUSAND DOLLARS ($150,000),,and hereby agree that if
you will advance such sum to me as hereinafter stated, I will assign to you an interest of fifty-one per cent. in all of said patents and inventions, and also in any patents or inventions which
I may hereafter secure or make having relation to or being useful
in any way in connection with electric lighting and wireless telegraphy or telephony. Such sum of ONE HUNDRED AND FIFTY THOUSAND
DOLLARS ($150,000.) is to be paid to me from time to time on my
submitting to you proper vouchers for expenditures made by myself in
connection with the subjects above mentioned.
            Understanding that you are prepared to accept this offer,
I herewith hand you assignments of an interest of fifty-one per
cent. in the patents designated upon the schedule hereto annexed,
and if, as and when, any other or further patents for inventions
relating to or useful in electric lighting, wireless telegraphy or
telephony shall be secured or made by me, I agree forthwith to
assign to you a like interest of fifty-one per cent. therein.
            Will you kindly confirm this understanding by letter.

                        Very respectfully yours,

Microfilm letters 14 and 15 appear to be earlier drafts of this letter, it looks as if
Tesla was unsure about the amount Morgan would invest. Morgan proposed to
do one tower first, the transatlantic one, for which he would fund 150,000
dollars.
According to an inflation calculator that would be roughly 4.5 million dollars in
2018.

New York March 4th, 1901.

Nikola Tesla, Esq.,

      46 East Houston Street, New York.

Dear Mr.Tesla:-

      I have talked with Mr.Morgan regarding the form of your letter, and it is satisfactery to him to have the sum of $150,000 filled in.

      Will you kindly have the letter completed in this respect and Mr. Morgan will confirm it at once.

                              Yours faithfully,

                              Charles Steele

JPM 0018 — transcript by Jeff Pearson

New York March 5th, 1901

46 & 48 East Houston St.

Mr. Charles Steele,
        23 Wall Street,
                New York City.

My Dear Mr. Steele:-
      Under inclosure I forward my formal letter to Mr.
Morgan, filled out and signed as requested!
      On this occasion I would renew my assurances of esteem
and gratitude and also express the hope, that in a time not distant
I may be able to prove myself worthy of the confidence he has
placed in me.
      Now, that all danger of conveying a wrong impression
to Mr. Morgan is removed by his kind acceptance of my proposal,
I would call to his attention, that I consider my fundamental
patents on methods and apparatus for the wireless transmission
of energy is the most valuable patents of modern times and, as
to my system of lighting, I am convinced, that it constitutes
one of the most important advances and is of enormous commercial
value.

Yours very sincerely,

Now that the deal is in, Tesla does mention the transmission of energy, but _not_ that his current proposal will implement this system.
He will do so later in a very unsuccessful attempt to save his project.

JPM 0019 — transcript by Jeff Pearson

New York, March 5th, 1901

Nikola Tesla, Esq.,

    46 East Houston Street, New York

Dear Sir:-

        I beg to acknowledge receipt of your letter of the 1st. instant together with assignments of an interest in various patents as shown upon the schedule and assignments handed me therewith, and to confirm the understanding therein expressed.

                        Very truly yours,

                            J Pierpont Morgan

JPM 0020 — transcript by Jeff Pearson

New York, March 7th, 1901

46 & 48 East Houston St.

J.P. Morgan, Esq.,

      23 Wall Street

        New York City

Dear Mr. Morgan:-

      I beg to acknowledge the receipt of your letter as confirming our understanding.

      It is one of my warning signs, that your investment would prove financially as profitable as it has been generous.

      With many thanks, believe me,

            Ever yours most faithfully

The deal is struck.

New York, August 8th, 1901

46 & 48 East Houston St.

J.P. Morgan, Esq.,

     23 Wall Street

        New York City.

Dear Mr. Morgan:-

        I beg to submit herewith a statement relative to the patents and applications partly assigned to you last March, which is self-explanatory.

        In a short time I expect to be able to report satisfactory progress along the two lines specifically mentioned in our agreement.

        Hoping in the interest of this country and the world at large for the conservation of your precious energies, I remain

        Yours most faithfully

Trouble begins

New York, Sep. 13th, 1901
46 & 48 East Houston Street

Mr. Stanford White,
        160 Fifth Ave.
                New York City.
My Dear Stanford:-
            I have not been half as dumbfounded by the news of the
shooting of the President as I have by the estimates submitted by you,
which, together with your kind letter of yesterday, I received last
night.
            One thing is certain: We cannot build that tower as
outlined.
            I cannot tell you how sorry I am, for my calculations show,
that with such a structure I could reach across the Pacific. Since last
night, I have thought carefully over the matter and have come to the
conclusion that the best plan will be to fall back on an older design
which I have made, involving the use of two and possibly three towers,
but much smaller. We would keep the design of the tower the same and
would only reduce the dimensions. It will probably be best to adopt a
design with two towers and a low central part for the machinery. I
shall make some calculations today and see how far I can reduce the
height without impairing materially the efficiency of the apparatus,
and will communicate with you as soon as practicable.
            Thanking you heartily for your friendly interest and
efforts on my behalf, I remain
                    Yours very sincerely,
                            N. Tesla

This letter is not part of the microfilm archive and was sourced elsewhere.
("Rare Notes" from Leland Anderson as mentioned earlier)
The initial design featured a tower of over 600 ft (183 m), which was reduced to
187 ft (57 m) in the final design.
For comparison: the Empire State Building stands 443.2 m (1,454 ft) tall while
the Eiffel Tower is 324 m (1,063 ft).

JPM 0022 — transcript by Jeff Pearson

New York, Oct., 13th, 1901
46 & 48 East Houston Str.

J.P. Morgan, Esq.,
          23 Wall Street,
                    New York City

Dear Mr. Morgan:-
          1. I respectfully apologize for disturbing you at a
time, when your mind must be filled with thoughts of a more seri-
ous nature than usual.
          2. My efforts are now centered upon that element of the
plant I am constructing, on which the effects, at distance chiefly
depend.
          3. The transmission of energy by my system is exception-
al in this respect, that a small increase in the sum invested in
the apparatus results in a very great increase in the performance
of the same.
          4. On the basis of many estimates which I have .....
made, you ..... it as not being far from true that, if the sum
..... in the element referred to be doubled, the apparatus will
..... to convey messages across the Pacific instead of only
across the Atlantic, and if the sum be tripled it will be fully
..... to convey signals to any point of the globe, no matter
what the distance.
          5. As to my .... of accomplishing this, I can only
assure you on my word that, were the technical difficulties a
thousand times greater, than they actually are, I would still be
certain of the result, if the limitations of capital were removed.
          6. I do not state this with the object of ..... you
to do anything to your example... generosity but merely to ac-
quaint you with a positive and unquestionable fact of incalculable
importance
          7. What I contemplate and what I can certainly accom-
plish, Mr. Morgan, is not a simple transmission of messages without
wires to great distances but it is in the transformation of the entire globe into a sentient being, as it were, which can feel in all
it's parts and through which thought may be flashed as through
a brain.
          8. Consider that from a single plant I can operate,
at the the same time, -not a hundred, or a thousand, or even a million,
but thousands of trillions - a practically infinite number - of
instruments, each costing no more than a few dollars, situated at
all parts of the globe.

9. Only two power plants, one located, say, on the old continent, and the other in America, can transmit more messages than millions of the present cables working continuously at full capacity, and this would be true even if their working capacity were increased a thousandfold.

10. You only need to realize this great truth to at once recognize, that as a means of conveying general knowledge the cables are absolutely and hopelessly doomed. They may be employed for special purposes, but not for the distribution of general news.

11. Here is a result, Mr. Morgan, of overwhelming humanitarion as well as financial importance, worthy of your energies, and one that would give you a new hold on the world, which has been already benefacted so much by your genius and activity.

12. Believe me, that however keen my satisfaction would be in completing this great work, an equal pleasure I would find in telling to the whole world that it is yours more that my own.

With expressions of admiration and gratitude believe me

Ever yours most faithfully,

Tesla's first response is to try and get some extra money. Tesla knew that his system would "advance humanity at least one century" and believed that Morgan would also want what is best for the whole world. Morgan ruled an empire worth more than one billion dollar, he would surely agree to a couple of thousands of dollars for the benefit of humanity...

J.P. Morgan & Co.

P.O. Box 3038

New York ———— Nov. 11 ———— 1.901.

Nikola Tesla, Esq.
46 East Houston St.
N.Y. City

Dear Sir,
    Mr. J.P. Morgan has
received your letter of
11ᵗʰ inst. with the
assignment of patents
mentioned therein.

Yours truly
Charles King
Private Secretary

Morgan however was more focussed on securing the patent-rights.

JPM 0025 - transcript by Jeff Pearson

46 & 48 East Houston Street
New York, Nov. 11th, 1901

J. P. Morgan, Esq.
            23 Wall Street
                    New York City

Dear Mr. Morgan:-
        Under inclosure please find assignment of patents which
were granted to me recently and which fall under the terms of our
agreement. They cover certain new departures in the transmission
of electrical energy through natural or artificial channels, and
it's utilization for various purposes. I am securing the rights in
Some of the foreign countries.
        Pardon me for trespassing on your valuable time in order
to give you, on this occasion, in the fewest words a clear idea of
the bearing of my patents relative to energy transmission so far
assigned.
        Any problem of this kind will present itself to you in
three aspects-first, the production of energy, second, it's
transmission, and third, its utilization at the distant point
        As to the first, my patents cover novel methods and ap-
paratus for the production of electrical effects of virtually unlimited power, not obtainable in any other ways here to fore known.
        In regard to the second my inventions cover the only
practical and economical method of transmission by conduction
through the earth with "tuned" circuits. They also cover new
methods of transmitting great amounts of energy for industrial
purposes without wire. The practical significance of my system re-
sides partially in the fact, that the effect transmitted diminishes
only in a simple ratio with distance, whereas in all other
systems it is reduced to the proportion to the square. To illustrate,
if the distance be increased one hundred fold, I get 1/100 of the
effect, while under the same conditions others can obtain, at the
very best, only 1/10,000 of the effect. This feature alone bars
all competition.
        In regard to the third, there are only two ways possible
of economically utilizing the energy transmitted for the production
of qualitative effects: either storing it in dynamic form as, for
instance, the energy of well-timed thrusts is stored in a pendulum
or by accumulating it in potential form as, for example, compressed
air is stored in a reservoir.

JPM 0026 — transcript by Jeff Pearson

In the patents assigned last spring I have embodied the first, and in some of those now assigned the latter principle. My rights on both are fundamental.

Referring particularly to telegraphy and telephony, I have still in the Patent Office two applications, one of which is allowed, while the other is about to be allowed. I have kept them back for a number of reasons. In one of them I describe and claim discoveries relating partially to the transmission of signals through the earth to any distance, no matter how great, and in the other a new principle which secures absolute privacy of messages and also enables the simultaneous transmission of any desired number of messages, up to many thousands, through the same channel be it the earth or a wire or cable. On this latter principle I have applied for patents in the chief foreign countries. I consider these inventions of extreme commercial importance.

Hoping that I shall soon be able to satisfy you that your generosity and confidence in me have not been misplaced, I remain,

Yours very respectfully

1902

JPM 0028 — transcript by Jeff Pearson

Morgan

Jan, 9 1902

... By the time my present plant is completed and I have made demonstra-
tion to your satisfaction, all this preparatory work would be done and,
should ..... you before the next winter the large plant could be put in
operation. Now, Mr. Morgan, am I backed by the greatest financier of all
times? and shall I lose great triumphs and an immense fortune because
I need a sum of money!! Is it not due to the honor of this country, that
it be identified with this achievement. Have I not contributed to it's
greatness and prestige and have my inventions not ....a revolutionary effect upon its industries. These are not my claims Mr. Morgan, only
my credentials.

This does not appear to be an actual letter. Not sure what this is.

38

JPM 0034 — transcript by Jeff Pearson

New York, Sep. 5th, 1902.

Dear Mr. Morgan:-

Hoping that you may now have a free moment to spare, I
venture to respectfully submit the following:

1. Under inclosure please find assignment of some foreign
patents, chiefly relating to an invention, which I have mentioned
in my earlier letters to you. In this country my fundamental
rights are likewise assured on this beautiful principle which permits of an indefinite cheapening and unlimited extension of telegraphy and telephony through natural and artificial channels.
Knowing your interest in scientific advances, I will attempt a
simple explanation. We are able to receive numberless distinct
impressions, because the controlling nerve fibers lend themselves
to innumerable combinations, and we can distinguish as individual
from all others by reason of a great many characteristic features
which in no other individual exist, or if so, not in that particular combination. In this reside both the essence and virtue of the
invention. The transmitter is not _primitively_ characterized by a
single note or peculiarity, as heretofore, but represents a very
complex and, therefore, unmistakable individuality, of which the
receiver is the exact counterpart, and only as such can it respond.
To go a step farther, I make the operation of the receiving instrument dependent not only of a great number of distinctive elements in combination, but also on their order of succession and,
if necessary, I go so far as to vary continuously the character of
the individual elements. These latter refinements I shall intro-
duce in the transmission to the most important centers of the
world. You can see, how impossible it will be to intercept my messages or to interfere with them. With proper facilities for manufacture and technical assistance it is practicable to prepare instruments which would enable the transmission of ten thousand simultaneous dispatches through either the Earth or a wire.

2. Some time ago, finding that my inventions in the wireless transmission of energy and production of electrical oscillations had been, boldly appropriated, I came to the conclusion, after
carefully considering every course I could adopt, that the only
way to fully protect myself was to develop apparatus of such power
as to enable me to control effectively the vibrations throughout
the globe. Now, if I had recognized this necessity earlier, I
would have gone to Niagara, and with the capital you have so gen-
erously advanced I could have accomplished this easily. But, unfortunately, my plans were already made and I could not change. I
endeavored once to explain this to you, much to my sorrow, as I
impressed you wrongly. Nothing remained then but to do the best I
could under the circumstances. My efforts will be in a large

-1-

measure rewarded, for by straining every part of my machinery to the utmost I shall be able to reach what I consider almost the maximum possible performance with the power available - a rate of energy delivery of ten millions of horse-power - more than twice that of the entire Falls of Niagara. Thus the waves generated by my transmitter will be the greatest spontaneous manifestation of energy on Earth. I believe I have told you in a previous letter, that in my system the strongest effect is produced at a point dia-metrically opposite the transmitter which, in this instance, is situated a few hundred miles off the western coast of Australia. The waves will be weakest in the countries which are at a dis-tance of about 6000 miles from here. It will be possible, however, to use the transmitter so that its effect will be _exactly_ equal over the whole surface of the globe, but then it will be compara-tively feeble, though the variation of potential will amount to fifty volts or more.

    3. It is now imperative to provide facilities for manu-facturing great numbers of receiving apparata. These will be in-expensive, and there is virtually no limit to the number which I can operate from my plant. They could be placed wherever desired and could be actuated simultaneously or separately. For this pur-pose it would be necessary to connect the plant by wires with the New York Telegraph and Cable Offices. At first, of course, the replies would have to come through artificial channels, but if agreements were entered into with Cable Companies the scheme would be very profitable to all concerned for the following reasons:

    a) revenue would be provided by charging for the wire-less messages; b) the existing lines would have more to do; c) the public would be benefitted by increased convenience and cheapening of the rates, and d) there would be no opposition and friction.

    4. Since your departure for Europe, Mr. Morgan, I have had time to reflect and to get a better knowledge of the importance and scope of your work, and I now see that you are no longer a man, but as a principle and that every spark of your vitality must be preserved for the good of your fellowmen. I have therefore given up the hope that you might aid me in establishing a manufacturing plant, which would enable me to reap the fruit of my labors of many years. But some ideas which I have not simply conceived -but worked out- are of such great consequence that I honestly believe them to deserve your attention. If you should wish it, I shall submit them to you, and in this hope I inclose a copy of an appli-cation[1] filed by me in the U. S. Patent Office on the 16th of May, 1900. A perusal of the paragraph on page 4 may convince you that what I contemplate are not unrealizable projects of a visionary, but rational undertakings of an engineer.

-2-

---

1 Patent 787412, "Art of Transmitting Electrical Energy through the Natural Mediums"

JPM 0036 — transcript by Jeff Pearson

  5. The .... of a factory with modern tools has some
.... to .... all the energy I could spare to this, and I
have originated a plan which has a fair chance of being put through.
I only need your kind approval, assuring you that whatever attitude
you may take, I have no greater desire than to prove myself worthy
of your confidence, and that to have had relations, however distant with so great and noble a man as you will ever be for me one
of the most gratifying experiences and most highly prized ....
times of my life.
  With these sentiments and awaiting your ...., I am
as ever
                        Yours most devotedly

Again Tesla hints at that his project includes the transmission of energy.

JPM 0037 — transcript by Jeff Pearson

New York, Sep. 17, 1902.

Dear Mr. Morgan:-

    Complying at the earliest possible moment with your wish, I forward under inclosure: 1) a brief descriptive of the inventions to be introduced, outlining plan proposed and accompanied by list of patents; 2) a few of my patent specifications with technical comments; and 3) copy of letter to the subscribers, to whom similar papers are to be submitted.

    I scarcely need say that none of those who are ready to take part in the enterprise and who are all people of high standing knows of your interest, and that all pertaining to this transaction will remain among friends, strictly private.

    As the Nikola Tesla Company, the owner of the patents, has incurred great expenses in perfecting the inventions, and as under this plan it would receive no reinbursement, it is proposed to sell a part of the bonds received by the Company for cash. I have thought it my duty to make the prospective subscribers understand this and no objection has been raised. This will enable me to refund the sum which you have in a generous spirit advanced, and as to your share in the whole property, it will be left entirely to yourself.

    Of my own interest I intend to give away one quarter to an associate, in whose ability and integrity I believe, and who is to unite all of his energies with mine, to bring to the greatest possible success this undertaking, in which our honor will be engaged.

    Hoping that there is nothing in this of which you would not approve, and assuring you of my deepest respect and gratitude, I remain, as ever

                    Yours most faithfully,

It looks as if Tesla had found additional investors to set up a factory for the receivers. Yet this has never materialized.

New York, Sept. 24th, *190*2

Nikola Tesla, Esq.,

    46 East Houston Street, New York.

Dear Mr. Tesla:-

    Will you kindly drop in and see me in the nest day or two about the matter concerning which you wrote Mr. Morgan.

                Yours very Truly,

                [signed] Charles Steele

JPM 0039 — transcript by Jeff Pearson

New York, October 21st, *190*2

Nikola Tesla, Esq.,

    46 East Houston Street, New York.

Dear Mr. Tesla:-

    I understand the plan of your proposed electrical company to be
as follows:

    The company is to acquire all your patents of which you spoke
to me and will be capitalized as follows:

Bonds,	$5,000,000
Preferred stock,	2,500,000
Common stock,	2,500,000

    It is intended that the company should go into the manufacture
of the various devices covered by your patents and for this purpose it is
proposed to raise a working capital of $2,500,000 by the sale of bonds
$2,500,000, preferred stock, $1,250,000, common stock $1,250,000. The
present Tesla Company, which includes yourself and associates, is to receive 40% of the new capitalization, via:

        $2,000,000 bonds
        1,000,000 preferred stock,
        1,000,000 common stock,

so that there will remain in the treasury of the new company,

        $500,000 unissued bonds,
        250,000 preferred stock,
        250,000 common stock.

    I understand your suggestion to be that of the securities which
go to yourself and associates, a certain amount will be used to provide

Nikola Tesla, Esq.,

cash required to reimburse the expenses which have been made to you and
the balance will be divided among yourself and associates.

I have discussed the suggestion with Mr. Morgan and it is in
substance agreeable to him. As to the amount which he should receive
out of the balance of securities remaining after providing for reinburse-
ment of advances, he does not understand the facts fully enough to decide.
If as he understands the principle patents to be acquired by the company
are those in which he hold an interest of 51% he thought that possibly
it would be fair to give him about one-third of the securities.

I should be glad to know what you think of this.

Yours faithfully,

[signed] Charles Steele

New York, April 8th, 1903

Dear Mr. Morgan:-

        I would respectfully report, that I have secured xxxxxx
U.S. Patents on the important principle repeatedly referred to in
my previous letters to you. There were four contestants in the
Patent Office, but all were late and my broad claims have been
granted in their entirety and without the slightest restriction.
This invention in itself makes it impossible for others to compete
with me in the lines of my present endeavors, as it secures to me
real privacy of the messages and permits the simultaneous transmission of virtually any desired number of them without interference. These improvements are, of course, being embodied in my
plant at Wardenclyffe, and will enable me to communicate intelligence to any part of the globe, irrespective of distance, with the
same precision and exclusiveness as through a private wire.
        I am ....fully.. ..ning the wireless speculators, who
since I began this work have quietly abandoned their processes
and appliances, which precluded the possibility of signalling to
any considerable distance and with precision, and have appropriated
my patents methods and apparatus, with which every blacksmith and
woodchopper can send reliably messages to great distances. As I
have told you, I have these people in my pocket. Their doings are
so manifest a departure from all rules of fairness, that they are
bound to collapse. My attorneys, Kerr, Page and Cooper assure me,
however, that I shall have no difficulty in shutting them up.
        It is my hope, Mr. Morgan, that nothing has occurred to
slight us in your appreciation, which has always filled me with a
sense of great honor and satisfaction. I recognize the fact, that
I should have resolved some practical results before expending so
much money, but under the circumstances it was impossible.
        First of all, You have raised great waves in the indu-
strial world and some have struck my little boat. Prices have gone
up in consequence -twice, perhaps three times higher than they
were. Then there were expensive delays, mostly at result of the
activities you have excited. The special machinery I needed I obtained only after nine months of promise, and I had literally to
beg for it. Then again I had a great number of patents to work
and to maintain. This took a considerable sum of money, for which
no provision whatever was made. Furthermore, as in every technical
project, unforeseen difficulties presented themselves, causing
trouble, delay and expense. But as I examine all my work today,

-1-

JPM 0043 — transcript by Jeff Pearson

I feel and can conscientiously assure you, that I could not have

come about it in a single feature. Improvements may be possible, but

they are beyond my knowledge and power. Two plants such as ....-
one here and one in the orient - can carry at least ten thousand
dollars a day.

Yours most faithfully
N. Tesla

JPM 0044 — transcript by Jeff Pearson

New York April 23, 1903

Dear Mr. Morgan:-

     My work is now in such shape that it can be completed in
three months. I would then be enabled to transmit at low cost absolutely reliable private dispatches to any part of the world, and
what this means I need not say.
     You have extended me a noble help at a time when Edison,
Marconi, Pupin, Fleming and many others openly ridiculed my undertaking and declared the success impossible. What has happened Mr.
Morgan, that just at present, when I am more than ever deserving
your support, You should hesitate? I am a century ahead in my art
and my patents cover the only practical means and methods for the
transmission and individualization of telegraphic and telephonic
messages without wires. With these patents you can control inventions which will create a revolution such as the world has not seen
before. Will you now let me go from door to door to humiliate myself, collect funds from some jew or promoter and have him par-
ticipate in this gratitude, which I feel for you?
     Hoping for your health and welfare, I remain, as ever,
                  Yours most faithfully,
                    N. Tesla

JPM 0045 — transcript by Jeff Pearson

J.P. Morgan & Co.

P.O. Box 3038

New York _____ July 3rd. _____ 1903.

Nicola Tesla, Esq.

    Waldorf-Astoria

        New York, N.Y.

Dear Sir:-

    Referring to your favor of even
date, addressed to our Mr. Morgan, will you
kindly call and see him on Monday, July 6th,
in reference to the subject therein referred to?

        Yours Truly,

        J.P.Morgan

Recorded at
THE PUBLIC TELEPHONE OFFICE
Time _____pm _____

To Mr. Tesla

Mr. J.P. Morgan has
written you, wishes
to see you at
his office on Monday
morning

JPM 0047 — transcript by Jeff Pearson

Copy.

July 3, 1903

Dear Mr. Morgan:-

Under inclosure I forward assignments of seven patents relating to the principles referred to in my previous letters. Nobody who respects these rights can guarantee secret and non-interferable messages.

I also sent a photographic view of my plant taken some time ago. The work has been carried on without interruption, but it is about as close to a oppage now as I can admit consistantly with my dignity. I have expended $130,000, which you have so generously invested and in addition $30,000, which I obtained by sale of a personal property, besides I have borrowed $10,000 from my bank and have a lot of bills to pay. Financially, I am in a dreadful fix. But if I can complete this work, I can readily show that by my wireless system power can be transmitted in any amount, to any desired distance and with high economy. Of the three hundred horsepower developed by my oscillator on Long Island two hundred and seventy-five, -perhaps a little more- can be recovered at the greatest distance in Australia. If I would have told you such as this before, you would have fired me out of your office. Now you see, Mr. Morgan, what I work for. I means a great industrial revolution. It will be something worthy of your attention, as I have always assured you. There is no incertitude about this, it is as absolute a fact as anything I know electricity can do. My patents confer a monopoly. Will you help me or let my great work - almost complete- go to pots?

Ever yours most faithfully
N. Tesla.

The date on this letter is probably wrong and should be July 10th.
(300 HP equals roughly 224 KW)
This is where Tesla kills his own project in an attempt to save it... Morgan is heavily invested in the AC distribution network and is certainly not interested in implementing a competitive system.
His answer is as expected:

JPM 0048 — transcript by Jeff Pearson

J.P. Morgan & Co.

P.O. Box 3038

New York _____ July 17 _____ 1,903.

N. Tesla, Esq,
          City,

Dear Sir,
I have received your
letter of 10th inst. and
in reply would say that
I should not feel
disposed at present
to make any further
advances.

          Yours truly
          J.P. Morgan

Tesla's Flashes Startling

That same Friday as Morgan wrote his reply the New York Sun published this article:

```
Tesla's Flashes Startling
```

But He Won't Tell What He Is Trying For at Wardenclyffe.

```
Wardenclyffe, L. I., July 16 — Natives hereabouts are intensely
interested in the nightly electrical display shown from the tall tower
and poles in the grounds where Nikola Tesla is conducting his
experiments in wireless telegraphy and telephony. All sorts of
lightning were flashed from the tall tower and poles last night. For a
time the air was filled with blinding streaks of electricity which
seemed to shoot off into the darkness on some mysterious errand. The
display continued until after midnight.
This morning workmen about the plant declined to say anything. It is
understood that Tesla has a well more than 200 feet deep driven beneath
the tower and that electrical waves are sent both underground and
through the air.
Mr. Tesla said last evening that the evidences of experiments which
have been seen by the people of Wardenclyffe, near which his laboratory
is located, have been going on for some time there, but that he had no
announcement to make at this time as to their character or the success
that had been achieved.
"It is true," said Mr. Tesla, "that some of them have had to do with
wireless telegraphy and that in addition to the tower and poles there
is a hole dug in the ground. This is 150 feet deep and is used in these
experiments. The people about there, had they been awake instead of
asleep, at other times would have seen even stranger things. Some day,
but not at this time, I shall make an announcement of something that I
never once dreamed of."
```

To me this and other reports of this event prove that Tesla could indeed accomplish what he set out to do. People reported seeing sparks near their feet as they walked on the street. This seems impossible without Earth resonance as he said he discovered in Colorado Springs on July 3rd 1899, and which I have experimentally confirmed in recent years.

JPM 0049 — transcript by Jeff Pearson

Sept. 13, 1903

Dear Mr Morgan:-

      Many years ago I was at your door with this invention, but I did not go in thinking that it would be useless.

      I have now another one, which is a hundred times more important and valuable. Help me to complete this work and you will see.

                  Ever yours most faithfully

                      N. Tesla

If Tesla could establish Earth resonance, which is evident to me, then he would certainly be able to send power worldwide. He would also be able to generate power, but he could not go to Morgan and say: "Look I have used up your money to build something entirely different than that what you paid for." Tesla had to complete the messaging part as well, a small little extra to which he had devoted only little attention compared to the whole rest of this system.

For his "art of individualization" he needed the tower to transmit multiple base frequencies and from his notes you can infer that he was struggling with implementing this in one single tower. Next, he would need to modulate the transmitted power so to let it carry messages. These were necessary before he could give a demo to his investor.

JPM 0050 — transcript by Jeff Pearson

The Waldorf-Astoria, New York,

Sept. 21, 1903

Dear Sir:-

I hope I did not impress you as asking for a generous contribution to science. It was simply a business proposition.

My last undertaking has returned more than two dozen times the original investment and this, in strong hand, should do better still.

I had the same difficulty in convincing the world of the practicability of my invention referred to in the editorial inclosed. The present one in incomparably more important and valuable.

Apologizing for the intrusion, I remain,

Very respectfully yours,

N. Tesla.

Tesla clearly does not understand what went wrong here and is desperately trying to figure that out. He is *so* close to success!

JPM 0051 — transcript by Jeff Pearson

New York, Sept. 24, 1903.

Dear Mr. Morgan:-

There is no day without such notices as the inclosed.
All this power can be easily transmitted by my new patented processes almost without loss and to any distance. If you will give
me your earnest support in this you can have a greater income from
these inventions than Rockefeller from his oil-wells. And you will
have at your mercy the cutthroats who are trying to undo your work
and get your royal mantle. I only need to complete this plant, Mr.
Morgan, the rest will take care of itself.

Yours most faithfully,

N. Tesla.

JPM 0052 — transcript by Ernst

The Waldorf-Astoria, New York

Oct. 13, 1903.

Dear Mr. Morgan,

     Please excuse me for troubling you at this time when, much as you may disdain the small doings of your enemies, you must be deeply pained to see the progress of your great economic ideas retarded. But it is precisely for this reason that I beg you now to hear me.

     I have never attempted, Mr. Morgan, to tell you even a hundredth of what can be readily accomplished by the use of certain principles I have discovered. If you will imagine that I have found the stone of the philosophers you will not be far from the truth. They will cause a revolution so great that almost all values and all human relations will be profoundly modified. These new developments do not concern any country in particular, but the whole world and they are in line with your efforts. The commercial possibilities they offer are simply infinite, and you are the only man today who possesses the genius and power to compel the universal adoption of these ideas and that is why I approached you two years ago. If I am only aided to reach the first result — the most insignificant of those to which I allude, yet very valuable — to girdle the globe with wireless messages, you will become interested and convinced. With my present experiences and knowledge of the wonderful appliances identified with my name this task is so easy, that it is for me no effort at all.

     I have had a long talk with Mr. Th. F. Ryan last night and he will see you in regard to providing the money still necessary. He is a great admirer and loyal friend of yours and for this reason as well as on account of his ability I am very anxious to enlist his cooperation. I have told him that one hundred thousand dollars will be sufficient to reach the first commercial results, which will pave the way to other and much greater successes. Knowing your generous spirit, I have told Mr. Ryan that any terms you may decide upon will be satisfactory to me.

Yours very respectfully,

N. Tesla.

Tesla contacted Thomas Fortune Ryan to see if he were interested in supplying the extra money that was required. While he certainly was, it did not work out. It appears that Morgan stopped him by creating an insurance scandal that involved Ryan.

JPM 0053 — transcript by Jeff Pearson

The Waldorf-Astoria, New York,

Dec. 7th 1903.

Dear Mr. Morgan:-

    Will you permit me to call this or any other evening at
your home and to bring a small instrument along to show you one or
two experiments with my "daylight"? I merely wish to have the
first honor, the shark who will probably come after me, will get
the contract for lighting your home.

                    Yours most faithfully

                         N. Tesla.

Tel. Waldorf.

copy.

The Waldorf-Astoria, New York,
Dec. 11th, 1903.

Dear Mr. Morgan:-
      Will you enable me to complete this work and show you,
that you have not made a mistake in giving me a checkbook to draw
on your honored house? It is my honest believe that I am a century
ahead in the fields in which I am working, and that is just the
trouble which confronts every pioneer. You see how your own advanced ideas are hampered.
      My enemies have been so successful in representing me as
a poet and visionary, that it is absolutely imperative for me to
put out something commercial without delay. If you will only help
me to do this, you will preserve a property of immense value.
      As regards the wireless project, I beg to call again to
your attention that my patents control <u>absolutely</u> all essential
features and that my work is in such shape, that whenever you
tell me to go ahead I shall girdle the globe within three months,
as surely as my name is Tesla. I have promised the St. Louis Exposition people to open the door of the Exposition with power transmitted from here. It is a great opportunity, Mr. Morgan. I can
easily do it, but if you do not aid me soon it will be too late.
Please think for a moment what this means for me. What I have told
you long ago has happened. My competitors have collapsed, since
wholesale appropriations as they have attempted do not go. Now
is the time to aid me, you know this better than anybody else.
                    Yours most faithfully,
                            N. Tesla.

WCO 0162 — transcript by Michelinho

Westinghouse Electric & Manufacturing Company,

Pittsburgh, Pa. December 12th, 1903.

OFFICE OF
L. A. OSBORNE
FOURTH VICE PRESIDENT.

Mr. Nikola Tesla,
46 East Houston Street,
New York City.

My dear Sir:-

In March or April, 1901 we directed our engineer in New York to deliver to you at East Houston Street one high tension testing outfit. You advised us at the time that you desired to borrow this for some months.

We now have very urgent need for this testing outfit and we would request that you deliver the same to our Mr. R. L. Wilson of our Construction department, 208 East 74th Street, New York.

We have advised Mr. Wilson to make arrangements with you regarding its return and we trust that it will be agreeable for you to do so at this time.

Yours truly,

L. A. OSBORNE
Fourth Vice President.

To make matters worse Westinghouse Co. is requesting the return of the high voltage equipment. Tesla makes an attempt to buy it, which for a moment looks hopeful.

Westinghouse Electric & Manufacturing Company,

Pittsburgh, Pa. Jan. 5th, 1904.

ADDRESS ALL COMMUNICATIONS TO
TO THE COMPANY AND REFER
TO THESE INITIALS   RCH

Mr. Nikola Tesla,
    #46 East Houston St.,
        New York City, N.Y.

Dear Sir:-

Your letter of Dec. 25th was duly received by us
and we have noted your request as to the billing to you of the
testing outfit which was loaned to you in March or April 1901.

Our price on this outfit to you will be One Thousand
Two Hundred Seventy-five Dollars ($1275), which is an especially
low confidential price.

Before we take any action regarding the billing of
this apparatus to you, we believe it would be well for you to send
us in your formal order covering this material, so that everything
may be in proper shape.

Kindly let us hear from you at the earliest date
possible.

Yours truly,

WESTINGHOUSE ELECTRIC & MFG. CO.
J. N. Koelord

RCH-EG

WCO 0165 — transcript by Michelinho

Westinghouse Electric & Manufacturing Company,

Pittsburgh, Pa. January 19th, 1904.

OFFICE OF
L. A. OSBORNE
FOURTH VICE PRESIDENT.

Mr. Nikola Tesla,
    46 East Houston Street,
        New York City.

Dear Sir:-

      Referring to your letter of January 11th in the matter of
the testing set; I have to advise you that in my absence your letter
was replied to and the impression given that it was our desire to sell
this piece of apparatus. As the matter stands, we are very much in
need of the testing set for work which we have in progress, as it is
impossible for us to make one and get it ready in time for the work
which we have in mind. It therefore becomes imperative that you return
the one which we have lent you and which you have had for over two years.

      If it is your desire to have a new one, we should prefer to
enter your order and make one and send it to you. We are much in
need at the present time of the one you have and will be greatly obliged
if you will send it to Mr. R. L. Wilson, care of Interborough Power
House, 59th Street, New York.

                Yours truly,

                    L. A. Osborne

                    Fourth Vice President.

But, no, it *had* to be returned and from this moment onwards the Magnifying Transmitter could probably no longer be powered.
Some believe that the 200 KW generator had been removed, but the foreclosure proceedings in 1922 report that this was still present in 1915.

Copy.

Dear Mr. Morgan:-

Since many years I have known one side of your character intimately. I believe that in my first approach to you I have given you evidence of this knowledge and I do so to-day. You have already put aside the money necessary to complete the work begun -in your thoughts, and that is as good as done. But I did not understand you as a business man until lately.

I have worked for results carrying with them a dignity and force such as to deserve your attention. What you wanted was simply a result. Now, Mr. Morgan, will you let me profit by this later knowledge and give me an opportunity to rehabilitate myself in your opinion as a business man? I need only a little ready capital to put some of my devices on the market. They will bring immediate returns. Will you make me an advance for this purpose and to preserve what I have already developed with so much pain, on any terms you choose?

Yours most faithfully,

(no date)

JPM 0056 — transcript by Ernst

Copy.

Dear Mr. Morgan:-

When this work is completed you will have something of incalculable value. As it now stands, I could be in three months in a position to establish communication with the chief cities of the world. This in itself would insure the realisation of the financial plan as outlined in Mr. Steele's letter to me of Oct. 21, 1902. In two months more I would be able to demonstrate that power can be transmitted by my wireless system, regardless of distance, without appreciable loss. Think only, Mr. Morgan, what this means in you hands. You can not permit that such a marvelous opportunity is lost, after what you have already done.

Would you be willing to advance the money if the Nikola Tesla Company were to authorize a bond issue for the amount. You could attach to this any condition you consider fair. My work is thorough, Mr. Morgan, it may seem long to complete, but I assure you that the ultimate results will be very gratifying to you.

(no date)

JPM 0057 — transcript by Jeff Pearson

Copy.
                New York, Jan 13, 1904

Dear Mr. Morgan:-
        1) The Canadian Niagara Co. will agree in writing to
furnish me ten thousand H.P. for twenty years without charge, if I
put up a plant there to transmit this power without wires to other
parts of the world. My system permits the transmission to be ef-
fected practically without loss. These people have great faith in
me. Besides furnishing the power they will take financial interest in my enterprise. With this splendid start I hope to realize
the project I submitted to you about a year and a half ago. As outlined then, I would use the energy, not for industrial purposes,
but for operating clocks, stock tickers and other apparatus, of
which there are millions now in use. On the average I shall require
not more than one tenth of one H.P. for each instrument. With ten
thousand H.P. I shall be able to offer a great convenience to the
whole world. I have calculated that with an xxxxxxxxxxx investment
of about half a million dollars from two to three millions can be
easily earned per annum. To illustrate one of the features, I
would start a manufacturing company there, which would simply turn
out clocks. These instruments will be extremely cheap, having
practically nothing but the hands and a couple of coils. They
could be readily produced for two dollars, and sold for, say, ten.
They will require no winding up, no attention at all and would in-
dicate <u>absolutely correct time</u>. They would be adopted everywhere
on land and sea.
        2) I have made preliminary arrangements for manufacture
of some of my inventions with a good manufacturing concern. (this
has nothing to do with the above.) They will produce and put together all purely machine parts and I shall have only the electrical parts to make. This reduces the capital which I need to manufacture to about one third.
        3) I am organizing an office in this City as Consulting
Engineer for the two fold purpose of, of facilitating all my commercial undertakings and arranging a permanent income. I have so far
never accepted a fee in my life. I am sure I can earn about fifty
thousand dollars a year in this capacity.
        Now, Mr. Morgan I want a small capital to put myself on
an earning basis. Will you help me on any terms you choose and
enable me to insure and develop a great property, which will ultimately yield hundredfold returns. Please do not do me an inju-
stice in believing me incapable simply because a certain sum of
money was not sufficient to carry out my undertaking. It is not my
fault that prices have gone up, that there have been delays, that
there is a panic. You may see that my work remains incompleted because of lack of funds, but you will <u>never see</u> that machinery

-1-

JPM 0058 — transcript by Jeff Pearson

which I construct not fulfill the purpose for which it was designed.

Today being my New Year, I take the opportunity to renew my heartiest wishes for your triumph over your enemies.

Yours Faithfully,

N. Tesla

Tesla tries everything, Morgan must be interested in the huge profits that his system would generate. But Morgan answers:

No more money!

JPM 0059 — transcript by Ernst

23 wall street   *Jan 1ˢᵗ 04*

My dear Sir,
    In reply to your note
I regret to say that I
should not be willing
to advance any further
amounts of money
as I have already told
you.   Of course I
wish you every success
in your undertaking.
            Yours very truly
                J.P. Morgan

N. Tesla, Esq,
Waldorf-Astoria

JPM 0060 — transcript by Ernst

The Waldorf-Astoria, New York,
Jan. 14th, 1904.

Dear Mr. Morgan:-
That was a nice letter to receive on my New Year! Had
you at least waited till to-day, bad news travel fast enough. You
wish me success! It is in your hands, how can you wish it?
We start on a proposition, everything duly calculated it
is financially frail. You engage in impossible operations, you
make me pay double, yes, make me wait ten months for machinery. On
the top of that you produce a panic. When after putting in all I
could scrape together I come to show you that I have done the best
that could be done you fire me out like an office boy and roar so
that you are heard six blocks away not a cent. It is spread all
over town, I am discredited, the laughing stock of my enemies.
It is just fourteen months that the constructive work on
my plant was stopped. If I would have been helped at that time
three months more with a good force of men would have completed it
and now it would be paying ten thousand dollars a day. More than
this, I would have secured contracts from governments for a number
of similar plants. I am the discoverer of the principles and the
inventor of all the essential devices and no one would have had the
slightest chance in a competition with me. You have favored the
schemers who have no knowledge or skill, but merely the cursing
sense of fraud to fool the world and to hurt my work more by
their incompetent attempts and far more than they ever could by
success.
Now when I have practically removed all obstacles skillfully put in my way and need only little more to save a great
property, which would pay you ten million dollars as surely as one
cent, you refuse me help in a trouble brought on by your own doings!
Twenty-five thousand dollars would enable me:
1) to start the manufacture of oscillators, which would
make my undertaking with you self-supporting and insure the ultimate success of my plans;
2) to put my light on the market;
3) to form a lighting company and realize the cash still
necessary to complete my plant;
4) the completion of this plant will put me in the position of carrying through the plan at Niagara;
5) my office will naturally facilitate all this and yield
a permanent income.
This would be to reach success in a slow and painful way.
If I had now what I need to complete the plant, that would be
different!

I am as anxious to success on your account as mine.
What a dreadful thing it would be to have the
papers come with your name
in red letters   It would be telegraphed all
over the globe. You may not care for it, Mr. Morgan, men are
like flies to you. But I would have to work five years to repair
the damage, if repairable at all. I have told you all. Please do
not write to refuse.
I am pained enough as it is.

                    Yours sorrowfully,
                              N. Tesla.

Tesla offers his professional services in the newspapers of January 1904:

I wish to announce that in connection with the commercial introduction
of my inventions I shall render professional services in the general
capacity of consulting electrician and engineer.
The near future, I expect with confidence, will be a witness of
revolutionary departures in the production, transformation and
transmission of energy, transportation, lighting, manufacture of
chemical compounds, telegraphy, telephony and other arts and industries.
In my opinion, these advances are certain to follow from the universal
adoption of high-potential and high-frequency currents and novel
regenerative processes of refrigeration to very low temperatures.
Much of the old apparatus will have to be improved, and much of the new
developed, and I believe that while furthering my own inventions, I
shall be more helpful in this evolution by placing at the disposal of
others the knowledge and experience I have gained.
Special attention will be given by me to the solution of problems
requiring both expert information and inventive resource — work coming
within the sphere of my constant training and predilection.
I shall undertake the experimental investigation and perfection of
ideas, methods and appliances, the devising of useful expedients and, in
particular, the design and construction of machinery for the attainment
of desired results.
Any task submitted to and accepted by me, will be carried out thoroughly
and conscientiously.

Nikola Tesla
Laboratory, Long Island, N. Y.
Residence, Waldorf, New York City.

JPM 0062 — transcript by Ernst

The Waldorf-Astoria, New York,
        Jan. 22nd, 1904.

Dear Mr. Morgan:-
        Are you going to leave me in a hole?!!
        I have made a thousand powerful enemies on your account,
because I have told them that xxxxxxxxxxx I value one of your shoe-
strings more than all of them.
        Do not grow old and .... as weak men do. You are
good for another twenty years, if you hold on to life, to your
people and young ideas.
        Could I not pledge you in .... my plant a while
for twenty-five thousand dollars I need to carry out the plan outlined in my last letter? If it is not worth that much, I am not an
engineer of world repute, but a chump.
        The better way, however, would be to enable me to com-
plete the plant at once. This would mean for you many millions of
dollars, and, what is more important still, a power which you could
use effectively. I hope that you never for a moment confound my
art with the incompetent efforts of my imitators. I could do better than any of them, if ninety-nine parts of me were paralyzed.
In a hundred years from now this country would give much for the
first honors of transmitting power without wires. It must be done
by my methods and apparatus and I should be aided to do it first
myself.

                        Yours most faithfully,
                            N. Tesla.

More and more desperate, Tesla begins to see the reality of this situation; that
Morgan is really not going to change his mind.

JPM 0063 — transcript by Ernst

Copy.   The Waldorf-Astoria, New York,

                              April 1rst, 1904.

Dear Mr. Morgan:-

        I have solved the greatest industrial problem which has
confronted humanity since ages. My success is absolutely certain.
I can deliver power in any desired amounts to any distance without
wire, in a practical and most economical manner. This will be of
incalculable consequence on the cost of necessities and commodities.
The prices of oil and coal in particular will be greatly affected,
merely by the moral effect of the first public demonstrations.

        I am tired of speaking to pusillanimous people who be-
come scared, when I ask them to invest five thousand dollars, and
get the diarrhoea when I call for ten. Will you aid me to complete
this great work? I have managed to advance it considerably. A
little more and I shall have practical results which will give you
the basis for a business of a magnitude such as the world has
never seen before.

                    Yours most faithfully,

                              N. Tesla.

JPM 0064 — transcript by Ernst

The Waldorf-Astoria, New York,

April 6th, 1904.

Dear Mr. Morgan:-

Please accept my best wishes for a happy journey.

The books I forward are great, but a thousand libraries
of such books will not be valued in a few years from now the article I enclose.

Yours most faithfully,

N. Tesla.

JPM 0065 — transcript by Ernst

New York, April 24, 1904.

Dear Mr. Morgan:-

        Have you ever read the book of Job? If you will put
my mind in place of his body you will find the sufferings accurately
described. I have put all the money I could scrape together in
this plant. With fifty thousand dollars more it is completed, and
I have an inmortal crown and an immense fortune.

                    Yours most faithfully,

                            N. Tesla.

Copy.

JPM 0066 — transcript by Ernst

T E S L A   L A B O R A T O R Y ,
Wardenclyffe, Long Island, N.Y.

July 22nd, 1904.

J. P. Morgan, Esq.,

New York City.

Dear Mr. Morgan:-

I hope the unfortunate misunderstanding, the cause of which I have been vainly trying to discover, will be removed and that you will recognize that my work is of the kind that passes into history and worthy of your support. My plant here, when completed, will enable you to talk from your office to any part of the world as clearly and distinctly as across your table, and it will make possible the transmission of telegraphic messages to all points of the globe with a speed and precision surpassing by far those practicable through wires, and its capacity of transmission will be greater than that of the entire cable system of the world combined. What is still needed can now be very closely estimated. Three months would suffice to complete the work, which has now been cruelly delayed for eighteen months. If you will aid me to the end, my country will be grateful to you.

Yours most faithfully,

JPM 0067 — transcript by Ernst

New York, Sept. 9, 1904.

Dear Mr. Morgan:-
        When you interested you..... it ...e .... that I
... .Despite of the unfortunate delay, I shall fulfil my promise to the letter, if you will
aid me to complete the work. Kindly .....   or that your ..... has
.... returns ...ly .... of the ....ture
from my original proposition. This plant .... will transmit
telegraphs and telephonic messages to any part of the world and
its earning power will be enormous, certainly not less than ten
thousand dollars a day, as I have already stated before, and besides, it will insure the universal adoption of my system. I am
also assured of contracts for several .... plants in England and
Russia. This is a work with which you will not be displeased to be
associated.

                Yours most faithfully,
                        N. Tesla.

        We know now that the a.......ments of transatlantic
messages were not founded on fact, but on this point I
to say more.

        Copy.

JPM 0068 — transcript by Ernst

New York, October 13th, 1904.

J.P. Morgan Esq.
        New York City.

Dear Mr. Morgan:-

I would beg you, in all earnestness, to peruse the following statement of facts which I have brought separately to your attention.

1. Five years ago, (as you may have gathered from from
my original announcement in the Century of June 1900, copy of a
patent specification filed May 16, 1900, and article in the Electrical World and Engineer of March 5, 1904) I succeeded in encircl-
ing the Earth with electrical waves. What gave to this result, far-
reaching in itself, a tremendous significance, was the observation
that in their passage, from Colorado Springs to the diametrically
opposite region of the globe and return, the waves suffered no per-
ceptible diminution of intensity, thus affording an absolute experimental evidence, that by my system power in unlimited amounts
can be transmitted, without wire, to any distance and, virtually,
without loss.

2. I recorded my discoveries in the Patent Office and
secured broad and uncontested rights in Patents, some of which I am
still keeping back, for reasons which it is unnecessary to explain.
When they appear they will create a profound impression.

3. I was, even then, firmly convinced that these advances
would prove of greater importance than the steam engine, the tele-
graph, the telephone and my multiphase motor combined, for they offered an ideal solution of the problem s of fuel, transportation,
and intelligence-transmission, in all their ramifications.

4. Desiring to obtain a support such as this work was
deserving, more for the good of the world than my own, I approached
you, naturally enough with the easily realizable project of establishing communication across the Atlantic, which required a
smaller investment.

5. I was fortunate to enlist your interest, but not
quite on the lines of my own suggestion. I contemplated the forma-
tion of one or two companies, to which all my inventions in wireless telegraphy and telephony and in my system of lighting were,
respectively, to be assigned, and proposed that you take fifty-one

J.P.M.,-2.

Percent. Of the stock (not fifty, as you yourself said in our first conversation, because then you would not control), the remainder to go to my Parent Company.   But when I received your formal letter it specified an interest of fifty-one percent. in _patents_ on these inventions.   That was different though my share was the same.   It was a simple sale. The terms were entirely immaterial to me and I said nothing, for fear of offending you.   Your have repeatedly referred to some stock and it is just possible, that a mistake was made, and that you intended to take exactly what I proposed, and what would have been, for many reasons, greatly to my advantage.

       6. Your participation called for a careful revision of my plans. I could not develope the business slowly in grocery shop fashion. I could not report yacht races or signal incoming steamers. There was no money in this.   This was no business for a man of your position and importance. Perhaps you have never fully appreciated the sense of this obligation.

7. When I discovered, rather accidentally,, that others, who openly cast ridicule on what I had undertaken and discredited my apparatus, were secretly employing it, evidently bent on the same task, I found myself confronted with wholly unforseen conditions.   How to meet them was the question. Of course I could not enjoin the infringers. In Canada, almost midway,,I had no rights. My patents on the art of individualization, insuring non-interference and non-interferability, were not as yet granted in England and the United States. Suppose I was anticipated in this invention?   Then I would have to rely on ordinary tuning. This was in a measure, satisfactory so long as I was alone, but shrewd competitors, with the advantage they had, could make me fall short, as the capital I had at disposal was only sufficient for two small plants. Once I failed with you in the first attempt, you would not listen to any other proposition. Once I lost your support I could not because of your personality and character of our agreement, interest anybody else, at least not for several years, until,the business would be developed and the commercial value of my patents recognized.   But there was one way, the only way, of meeting every possible emergency, and making the ultimate success perfectly certain.

       8. Here I must add a purely explanatory paragraph. Suppose a plants is constructed capable of sending signals within a given radius, and consider an extension to twice this distance. The area being then four times as large the returns will be, roughly, fourfold on account of this alone. The messages, however, will become more valuable. Approximately computed, the average price will be tripled. This means that a plant with a radius of activity twice as large will earn twelve times as much.   But it will cost

J. P. M.,-3.

Scarcely twice as much. Hence in investing a certain sum destined
for two small plants into a single one, the earnings will be six-
fold increased. The greater the distance the greater the gain until, when the plant can transmit signals to the uttermost confines
of the Earth, its earning power becomes, so to speak, unlimited.

　　　　9. The way to do was to construct such a plant. It
would yield the greatest returns, not only for the reasons just
mentioned, but also because every other plant erected anywhere in
the world, by anyone, was sure to be turned into a source of income
It would give the greatest force to my Patents and insure a mono-
poly. It would make certain the acceptance of my system by all
governments. It discounted in advance all possible drawbacks, as
anticipation of the results by thee trespassers of my rights and
delay. It offered possibilities for a business on a large, digni-
fied scale, commensurate with your position in life and mine as a
pioneer in this art, who has originated all its essential principles.

10. The practicability of such an undertaking I had already demonstrated in Colorado, but to make those feeble effects,
barely detectable by delicate instruments, commercially available
all over the Earth, required a very large sum of money. You had
told me from the outset that I should not ask for more, but the
work was of such transcending importance and it was of such enormous value in your hands, that I undertook to explain to you the
state of things on your first return from abroad. You seemed to
misunderstand me. That was most unfortunate. Had I obtained your
hearing, your enemies would not had succeeded in inflicting you
injuries, for the first motor or lamp operated across the Pacific,
would have delivered them in your power. To achieve a great result is one thing, to achieve it at the right moment is another.
That favorable moment is gone forever. Your popularity has suffered, the moral force of my work has been weakened by delay, the
audacious schemers who have dared to fool the crowned heads of
Europe, the President of the United States, and even His Holiness
the Pope, have discredited the art by incompetent attempts and
spoiled the public by false promises which it cannot distinguish
from those sure of fulfillment, based on knowledge and skill and
legitimate right.   That is what pains me the most.

11. Still, in spite of all this, Mr. Morgan, I can
realize what I have held out to you when you yourself said to me
that "you had no doubt". I know you must be sceptical about get-
ting hundredfold returns, but if you will help me to the end you
will soon see that my judgement is true. Once my first plant is
completed I can place a dozen of such at once. I do not need to
wait for returns from subscribers. There are one thousand million

J. P. M.,-4.

Dollars invested in submarine cables alone.   This immense property is threatened with destruction because just as soon as people find that messages for, say, five cents a word can be transmitted to any distance, nothing will stop the demand for the cheaper and quicker means of communication.   The investment in cables is too large to pay on this low basis and the only chance the Companies have is to take hold of the new advances.   My patents control every essential element of the art.   They are impregnable.   In your hands, and backed by these great results, they should be of enormous value.

   12.   My work is now so far advanced and could be finished quickly.   I have expended about $250,000 in all and a much smaller sum separates me from a great triumph.   If you have lost faith in me have you not someone in whose knowledge and ability you have greater confidence than in mine, and to whom I could explain? Seventy-five thousand dollars would certainly complete the plant and then I would have no difficulty whatever in getting all the capital necessary for the further commercial expansion.

   13.   Since a year, Mr. Morgan, there has been hardly a night when my pillow was not bathed in tears, but you must not think me a weak man for that.   I am perfectly sure to finish my task, come what may.   I am only sorry that after mastering all the difficulties which seemed insuperable, and acquiring a special knowledge and ability which I now alone possess, and which, if ap-plied effectively, would advance the world a century, I must see my work delayed.

   In the hope of hearing from you favorably, I remain,

   Yours most faithfully,

Here Tesla explains why his two smaller towers merged into one. I still find it strange, though, for once you establish Earth-resonance it can be felt everywhere, except perhaps for some nodal lines. I think Tesla did not have messaging on his mind at all, but was focussed on power distribution. The possibility of messaging being a mere by-product.

JPM 0072 — transcript by Ernst

23 WALL STREET

New York, Oct.15th,1904.

Mr. Nicola Tesla,
    Waldorf-Astoria,
            New York City.

Dear Sir:-

Referring to your letter of
13th October, Mr. J. P. Morgan wishes
me to inform you that it will be impossible for him to do anything more in
the matter.

Yours truly,

Private Secretary.

JPM 0073 — transcript by Ernst

New York, Oct. 17th, 1904.

Dear Mr. Morgan:-

    You are a man like Bismarck, great but uncontrollable.
I wrote pruposely last week, hoping that your recent association
might have rendered you more susceptible to a softer influence.
But you are no Christian at all, you are a fanatic .us..lman. Once
you say no, come what may, it is no. May the gravitation repel
instead of attract, may right become wrong, every consideration, no
matter what it be, must founder on the rock of your brutal resolve.

    It is incredible. A year and a half ago I could have
delivered a lecture ...., which would have been listened to by all
the academics of the world, in the tone of my voice! That would
have been the time to thank you.
    You let me struggle on, weakened by shrewd enemies, dis-
heartened, by ....y friends financially .yka.ated, trying to
overcome obstacles which you yourself have piled up before me.
    I know, Rankine told me what a time they had in placing
the Niagara bonds. And what a time must I have?! "If this is a
good thing, why does not Morgan see you through?!" "Morgan is the
very last man to let a good thing go!" .. it has been going on for
two years. I advance, but how? Like a man swimming against a
stream that carries him down.
    Will you not listen to anything at all? Are you to let
me perhaps succomb, loose an immortal crown, will you let a property of immense value be depreciated, let it be said that your own
judgement was defective, simply because you had once said no. Can

not I make you a new proposition to overcome the difficulty? I

tell you I shall return your money a hundredfold.

        Yours faithfully,

            N. Tesla.

JPM 0074 — transcript by Ernst

The Waldorf-Astoria, New York

Nov. 5, 1904.

Dear Mr. Morgan:-

The inclosed bears out my statement made to you over a year and a half ago. The old plant has never worked beyond a few hundred miles. Apart of imperfection of the apparatus inside there were four defects, each of which was fatal to success. It does not seem probable that the new plant will do much better, for these faults were of a widely different nature and difficult to discover.

As to the remedies, I have protected myself in applications filed 1900 — 1902, still in the office.

Yours faithfully,

N. Tesla.

This appears to be a very strange letter. If this refers to the Wardenclyffe tower and the Colorado Springs station, as it appears to, then this would not exactly convince anyone to invest (additional) money in the project. On the other hand Tesla mentions he knows how to solve the problems… But has not implemented these remedies in the new plant??
My best guess is that it refers to the work of others.

JPM 0075 — transcript by Ernst

The Waldorf-Astoria, New York,   Nov. 18th, 1904.

Dear Mr. Morgan:-
        The inclosed written by the ablest man in the field of
electrical transmission by my system reminds me of the difficulties
I had in convincing people of the value of that invention. What
I have now is immensely more valuable and consequential.
                        Yours faithfully,
                                N. Tesla.

        (Sent: Editorial by C. F. Scott from Electric Club
Journal, October, 1904.)

        Copy.

JPM 0076 — transcript by Ernst

Copy.

New York, Dec. 16, 1904.

Dear Mr. Morgan:-
        1.If you will advance me one hundred thousand dollars,
which will enable me to complete my plant and instal some telegraphic and telephonic receivers on a few of the most important points
in the world, I will return on your investment of $250,000 twenty-
five millions. To attain this result it is necessary that the
work be attacked at once.
        2.   Assuming that you are not willing to do this, if you
will advance me fifty-thousand , which will suffice to finish the
indispensable parts, make everything perfectly fire-proof, as I
have planned and take out an insurance, my ultimate success will be
rendered certain, and although you may have to wait long, I shall
still and without fail carry out my promise of returning to you one
hundred times the sum invested.
        3. If you do not want to do this, the only one thing re-
mains. You release me of all obligation, give me back my assignments, and consider the sum you have invested as a generous contribution, leaving it all to my integrity and ability to work out the
best results for you and for myself. In this case I intend to go
on a lecture tour, which I believe will give me enough money to
complete the plant. Once I am so far it would not take me more than
a week to get a few millions in Wall Street. I hope sincerely,
that you will not resolve yourself to this last course. Believe
me, you would give me great pain.

                          Yours faithfully,
                                    N. Tesla.

This is where Morgan's attitude towards this project becomes more shady. He
already kept other potential investors away, most probably by requiring full
reimbursement without giving up the patent-rights.
Now, with his answer it becomes clear that he does not want the project to
proceed.

JPM 0077 — transcript by Jeff Pearson

23 WALL STREET   *N.Y Dec 17/04.*

N. Tesla, Esq,
      Waldorf Astoria,
           City.

Dear Sir:-

      I have received your note of 16th inst and in reply would state that I am _not_ willing to advance you anymore money, as I have frequently told you.

      As to your third proposition I am not prepared to accept this either. I have made and carried out with

*You in good faith a
Contract, and, having
performed my part,
it is not unreasonable
that I expect you to
carry out yours.*

> *Yours very truly,*
>
> *J. Pierpont Morgan*
>
> *per C. W. King*
>
> *Private Secretary*

*(Letter dictated by
Mr JP Morgan.)*

JPM 0079 — transcript by Ernst

New York, Dec. 18th, 1904.

Dear Mr. Morgan:-

 To-day is my patron saint, who has always stood by me.

 Will you let me complete the important work and tell the
world, in a way which will be forever recorded in history, that I
have to thank only to the generosity of a great man and my own
efforts?

 I am fully aware, Mr. Morgan, that your hands are full and
that you are perhaps grieved. But permit me to tell you that in
the eyes of those who see far you have shown yourself greater and
stronger than ever before. (those that show you believe in your
complete triumph?)

 Yours most faithfully,

 N. Tesla.

JPM 0080 — transcript by Ernst

Copy.

New York, Dec. 19, 1904.

Mr. J. P. Morgan,
        New York City.
Sir:-
        Owing to a habit contracted long ago, in defiance of su-
perstition, I prefer to make important communications on Fridays
and the 13th of each month, but my house is afire and I have not an
hour to waste.
        I knew that you would refuse.. What chance have I to land
the biggest Wall Street monster with the soul's spider thread!
        Your letter reached me just on the day of my patron saint
- the greatest of all — St. Nikola. There was a silent agreement
between St. Nikola and myself that we would stick to each other.
He did well for a time, but during the last three years he has for-
gotten me, - as you have.
        You say that you have fulfilled your contract with me.
You have not.
        I came to you to enlist your genius and power, not be-
cause of money. You should know that I have honored you in so doing as much as I have honored myself. You are a big man, but your
work is wrought in passing form, mine is immortal. I came to you
with the greatest invention of all times. I have more original
creations named after me than any other man that has gone before,
not excepting Archimedes and Galileo — the giants of invention.
Six thousand million dollars are invested in enterprises based on
my discoveries in the United States to-day. I could draw on you at
sight for a million dollars, if you were the Pierpont Morgan of old.
        When we entered our contract I furnished 1) patent-rights,
2) my ability as engineer and electrician, 3) my good will. You
were to furnish 1) money, 2) your business ability, 3) your good
will. I assigned patent-rights which, in the worst case, are worth
ten times your cash investment. You advanced the money, true, but
even this first clause of our contract was violated. There was a
delay of two months in furnishing the last $50,000 — a delay which
was fatal.
        I complied conscientiously with the second and third ob-
ligations. You ignored yours deliberately. Not only this, but you
discredited me.
        There is only one way to do, Mr. Morgan. Give me the
money to finish a great work, which will advance the world a century and reflect honor on all that come after you. Or else, make me
a present and let me work out my salvation. Your interest is sa-
cred to me and my hearty wishes for your happiness and welfare will
always be with you.
                        Faithfully yours,
                                N. Tesla.

JPM 0081 — transcript by Jeff Pearson

(Letterhead) *New York,* January 6th, 1905

Nikola Tesla, Esq.,

        Waldorf-Astoria

            New York, N.Y.

Dear Sir:-

     We enclose herewith receipted bill for Exchange ordered by you and also check for difference.

               Your truly,

               (signature)

Enclosures 2.

JPM 0082 — transcript by Jeff Pearson

Copy.

New York, Feb. 19th, 1905

Dear Mr. Morgan:-
    Let me tell you once more.
    I have perfected the greatest invention of all times -
the transmission of electrical energy without wires to any distance,
a work which has consumed ten years of my life. It is the long
sought stone of the philosophers. I need but to complete the plant
I have constructed and in one bound humanity will advance centu-
ries. I am the only man on this earth to-day, who has the peculiar
knowledge and ability to achieve this wonder and another one may
not come in a hundred years. There has been a long and painful
delay. My nerves are hot of iron, and all this knowledge and ability may be lost to the world. Please do not delay further to act
either one way or the other. Help me to complete this work or else
remove the obstacles in my path.
    I was heartily glad to see you in such splendid health
yesterday. You are good for another twenty years of active life.

Faithfully yours,

N. Tesla

JPM 0083 — transcript by Jeff Pearson

Copy.

New York, Feb. 27th, 1905.

Mr. J.P. Morgan,

23 Wall Str., New York.

Dear Mr. Morgan:-

Understanding that you are to sail the day after to-morrow I would earnestly beg you not to leave without some understanding offering me a chance to complete soon the work which, but for unfortunate circumstances, would have been finished three years ago. The attention of the whole world is turned upon it and something must be done for your as well as for my sake.

There are a number of ways which might lead to quick results without further expenditures on your part. If you will kindly accord me a few moments at your convenience, for the purpose of choosing a plan, I believe that in a time not very distant you would have every reason to be satisfied.

Yours Faithfully,

Copy

signed N. Tesla

JPM 0084 — transcript by Fabrice de Gaudemar

New York, August 11th, 1905

Mr. J. P. Morgan

    23 Wall Street, New York

Dear Sir;-

    Under enclosure, please find assignment of a patent
recently granted to me on certain discoveries to which I
have referred in my previous correspondence. In view of
this, the specification I forward might possibly be of
interest to you. Rights have been reserved in three foreign
countries.

    On this occasion I take liberty to inclose
a letter addressed to you shortly before your departure
for Europe. My Secretary who was its bearer and in your
presence, seems to be under the impression that it was
miscarried.

                Yours faithfully,

94

JPM 0085 — transcript by Fabrice de Gaudemar

New York, Sep 29th, 1905.

Copy.

Mr. J. P. Morgan,

    23 Wall Str., New York.

Dear Sir:-
        I have developed conjointly with another well-known
practical engineer a novel system for the propulsion of vessels, in
all important respects superior to any now in use. We eliminate
virtually all moving machinery, reducing the whole apparatus to
simple, stationary elements, apply the energy of the fuel more di-rectly
and with a minimum loss and complication. This means reduced cost of construction, much greater speed, cruising radius and
carrying capacity. The system is readily applicable to any ex-
isting vessel at small expense and without material alteration,
except for the removal of engine, boilers, condensers, smoke-
stacks, etc.
        I have never forgotten your generous help in the develop-
ment of inventions which are so far in advance of time, that they
have as yet brought no commercial returns. This is a peculiarly
timely, strictly commercial engineering proposition, which should
be of exceptional value in the present unsettled state of the art,
and it is herewith respectfully submitted to you for refusal or
acceptance on your own terms.

                        Yours faithfully,
                            (signed) N. Tesla

Tesla is trying other ways to attract funds. The interesting thing here is that most
of this work consists of applying his electrical discoveries to other fields. The
root cause of electrical effects is found, according to Tesla, in a gaseous medium.
Laws that apply to electricity should therefore also apply to any other gaseous
medium and vice versa.
This believe is clearly demonstrated in Tesla's work and patents after 1901.

JPM 0086 — transcript by Fabrice de Gaudemar

23 WALL STREET [paper header]

New York, Sept. 30,1905

Nikola Tesla, Esq.,
        Waldorf-Astoria,
                New York City

Dear Sir:-

        Referring to your letter of
29th inst., Mr. Morgan has carefully read
same but regrets he cannot do anything in
regard to the matter which is entirely
out of his line.

                Yours truly,

                        [signed]Brokings

                        Private Secretary.

JPM 0087 — transcript by Fabrice de Gaudemar

New York, December 13th, 1905.

Mr. J.Pierpont Morgan,

23 Wall Street, City

Dear Sir:-

Under enclosure please find Assignment of two patents
which have just been granted to me and come within the terms of
our agreement.

These patents describe an important discovery which I
made in Colorado in 1899. Ordinarily the Earth responds very
feebly and only locally to electrical disturbances much like the
Ocean to a stone thrown into it. But there are certain waves to
which the Earth is sympathetically responsive and when these are
impressed upon it the globe as a whole responds very vigorously.

I take liberty, on this occasion, to tell you that I
have just recovered from a severe illness which has brought more vividly
to my mind the unfortunate state of affairs confronting
me since three years. Just fourteen years ago I undertook to
solve the problem of transmitting electrical energy without wires
and in 1899, after spending a large private fortune, I finally
succeeded in achieving what seemed impossible. As you will kindly
remember, I approached you with a very easy and simple proposition
of establishing telegraphic communication across the Atlantic, in
the hope that I might later interest you in the transmission of
power for industrial uses. Circumstances which we did not fore-
see made the original proposition impracticable and I adopted a
plan which would make my success absolutely sure, irrespective
of what might be developed by others and which would lay the

foundation to a development commanding universal respect. You
never advised me with one word and I did the best I could myself.
The result has been that for want of a certain sum of money my
work has remained, up to this day, unfinished. But despite all
the unfortunate delay I could still return twenty times your investment if I could complete my plant, for then I would be able
to send messages in type, in handwriting and by telephone to any
part of the world. More than this, I could establish clearly
the fact that with my system, energy in unlimited amount can be
transmitted practically without any loss to any distance.

As you will see, Mr. Morgan, all depends on my being
able to complete this work. To do this, more money is required.
Whoever advances this capital must receive a fair interest.
Consequently if there is much a person to be found it all depends
on you for, as fas as I am concerned I am satisfied with the
smallest interest. I believe that I have found a man generous
enough to advance One hundred Thousand Dollars. It is Mr. Frick,
whom you well know. What has passed between us has impressed me
with the feeling that he will certainly do it if I can only frame
some kind of a proposition in harmony with you. I beg you to
consider what the completion of this work means to me and to the
world and act accordingly in your noble magnanimous way in which
you have already assisted me and for which I am profoundly grateful.

Yours very respectfully,

Now Tesla has found Henry Clay Frick willing to supply the remaining sum.
Frick is one of the richest men in the world at his time, but it would be foolish to
go against Morgan.

JPM 0089 — transcript by Fabrice de Gaudemar

23 WALL STREET　　　　　*Dec 14/05*

My dear Mr Tesla,

I have received your letter of 13th Dec and in reply would state that I am not willing to invest any more money in the enterprise. I should be very glad if Mr. Frick would join you. You could have no better associate and I should be very glad to work with Mr. Frick in the matter putting in what I have against his $100,000 to which you allude.

N. Tesla by
Waldorf Astoria
NY City

Yours very truly
(signed) J.P.Morgan

This looks like Morgan approves of this proposal, but it did not materialize. What I read in other sources is that Morgan required reimbursement of his 150,000 dollars without giving up the patent-rights. Other than that Morgan also started to oppose other businesses that Frick had an interest in. Thus sending a signal to all possible investors to stay away.

JPM 0090 — transcript by Ernst

Dec. 15 1905.

Dear Mr. Morgan,

I thank you a thousand times for your very kind letter received this morning. You could not have given me more pleasure if you had included your chec for the amount I need. Pardon me for this annoyance. I want him

This handwritten letter spans 6 pages, the rest of which is illegible.

A Nervous breakdown

JPM 0097 — transcript by Fabrice de Gaudemar

C O P Y.

Feb. 15, 1906.

Dear Mr. Morgan,

We are committee in an enterprise which <u>must</u> be carried to success for you as well as my own sake. It <u>will be</u> if you will give me a reasonable foundation to build upon, to which I am justily entitled.

Will you agree, by letter, that in exchange for your patent interest you will take one-third of the stocks and bonds in companies which I shall organize for the expoundations of my inventions in telegraphy and telephoning without wires and lighting by my system? I suggest one third for you only because it is essential for me to show others, that all are coming in on the same basis. If you will do this before sailing you will give me a chance of putting my delayed projects through. You need not fear that anybody, with whom you would not care to associate, will get in. Out Caesar out Nihil is my principle.   Please do not spoil the letter by an unnecessary reference to your unwillingness of furnishing more money. The whole town knows it.

                    Yours faithfully,

JPM 0098 — transcript by Fabrice de Gaudemar

COPY.

New York, Feb 16th, 1906

N.Tesla, Esq.,

   Waldorf-Astoria,

      City.

Dear Sir:-

     I would say that I am quite prepared to take one third of any securities of any Company that you may be interested in, or under your control, as full settlement, so far as that particular Company is concerned, for my loan, provided there is secured other capital adequate for the purpose of carrying out the business.

          Yours truly,

        (Signed) J. Pierpont Morgan.

In other words: I am quite willing to accept additional securities, but I will hold on to those patent-rights.

JPM 0099 — transcript by Fabrice de Gaudemar

C O P Y.

Feb. 16, 1906.

Dear Mr. King,

My letter to Mr. Morgan yesterday was written in a hurry and needs amplification.

As you know Mr. Morgan owns an undivided patent interest in some of my inventions. He could go ahead without me, and so could I without him. This destroys the value of both the separate rights. To enlist new capital it is essential that our interests be merged. A statement from Mr. Morgan to this effect is therefore indispensible. Both the necessity and justice of this will certainly not escape him.

In proposing a third for him I can only hope to get for my parent Company--- the original owner of these franchises--- sixteen per cent, as the control will have to be surrendered to the new capital. My personal interest will consequently be small.

Will you kindly bring this before Mr. Morgan at a favorable moment today. Should he sail without doing any thing in this matter he would make me absolutely helpless.

Although one of Mr. Morgan's partners has told me that he is disgusted with me, I have never believed it. On the contrary I have always felt sure of his sympathy and feel all the more so now after passing through misfortunes the like of which can scarcely be found in history.

In granting my requests he will give me a chance. Only a slim one, unfortunately, for a project abandoned by Mr. Morgan is as dead as a door nail.

Yours very truly, N. Tesla.

Now a series of handwritten notes follows such as this one. Many parts are crossed out and most of it is completely illegible.

JPM 0116

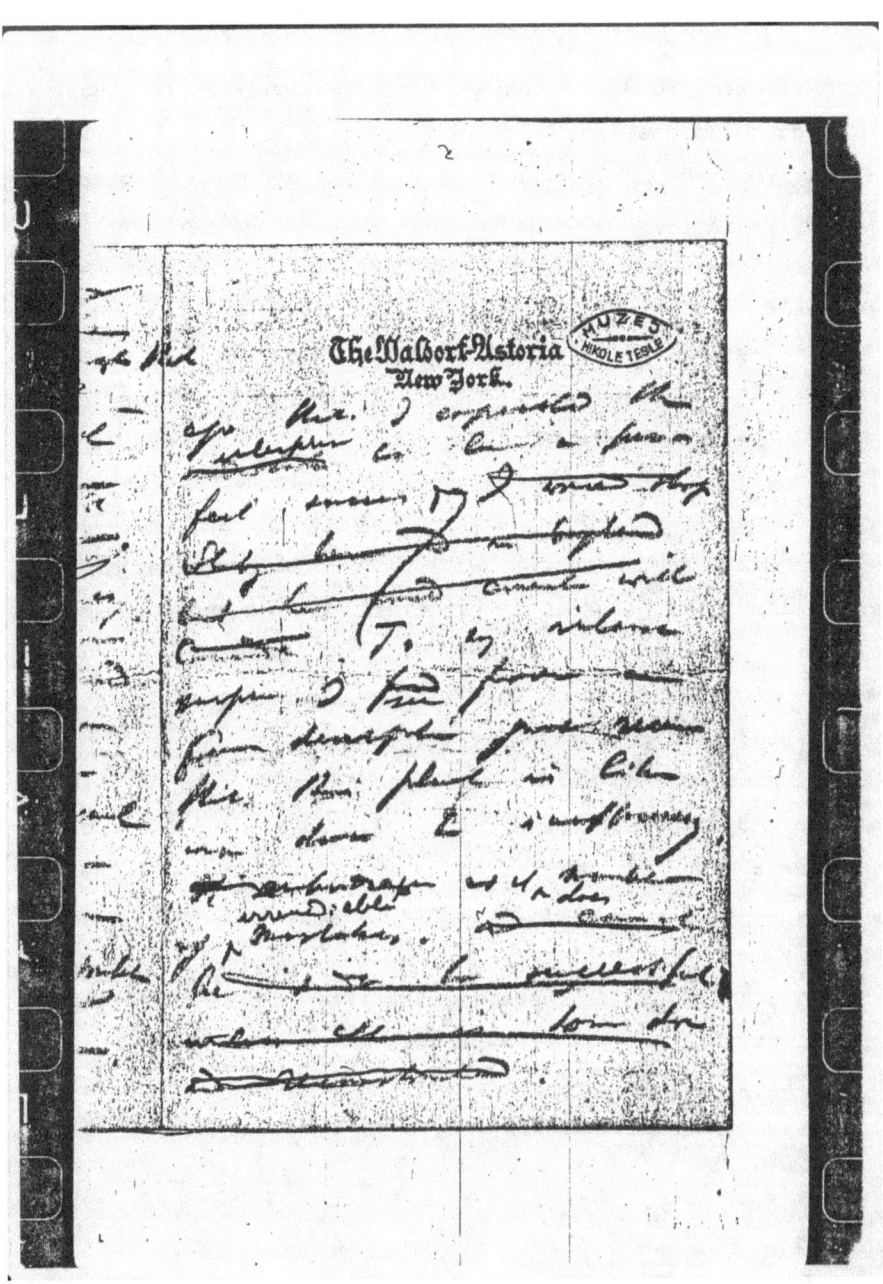

According to Marc J. Seifer, known for his biography of Nikola Tesla and for his handwriting analysis, 1906 was a most terrible year for Tesla. Not only did he suffer fierce opposition from Morgan from whom he expected generous help, but also two of his dear friends died around that time, William Birch Rankine on September 30th, 1905, and Stanford White on June 25th, 1906.

Seifer beliefs Tesla suffered a nervous breakdown and finds confirmation in a letter of George Scherff, his secretary.

GS 0306 — transcript by Michelinho

T E S L A   L A B O R A T O R Y ,
Wardenclyffe, Long Island, N.Y.

April 10th, 1906.

Dear Mr. Tesla:-

I have received your letter and are very glad to know, that you are vanquishing your illness. I have scarcely ever seen you so out of sorts as last Sunday and I was frightened.

The car of coal arrived this morning; Hawkins and Peter are unloading it.

The paint, brushes, brass polish and a box of tools have also come.   I have heard nothing as yet from Koven Bros. regarding the tank and am writing them.

I found yesterday that the paper in the new condensers had sunk down considerably with its own weight and built them up a little higher.   Six hundred sheets of tinfoil went in easily.

Do you wish to have the charging coil rewound with #23 wire?

I wrote to Eisenmann, but have not yet received an answer.

You asked me the other day about the initials of Mr. M's secretary.   His name is C. W. King.

Later that year Scherff notifies Tesla that he needs to find other employment as Tesla can no longer pay his salary.

They part as good friends and remain so for many years.

JPM 0145 — transcript by Jeff Pearson

COPY.
Waldorf-Astoria
New York, Dec. 18, 1907.

Mr. W. C. King,
25 Wall Street,
New York City, New York.

My dear Sir:
    Acting on your suggestion I made my letter to Mr.
Morgan with reference to patents as short as possible, but
that made it necessarily incomplete. As I am anxious no to
disturb him I would ask you to kindly bring the following, at
the earliest propitious moment to his attention.

    The value of the patents is not so much dependent on the
commercial success of the original invention as on the general
development of the art. Whoever succeeds commercially is bound
to put his business under the protection of the controlling
rights for the simple reason that he can make more money in
this way than by fighting. Besides, it is a matter of principle.
A competent business-man is not disposed to devote himself to an
undertaking with the sword of Damocles constantly above his head.

    Now I have in the U.S. Patent Office since four years
several inventions which should be protected in the foreign
countries without delay. They are of the greatest importance
and value. It would be difficult to explain them in a short
letter, but just to convey an idea ot one suppose that all
boilers heretofore made were of a material full of holes through
which steam would escape so that the most furious fire would
not raise but a feeble pressure. My discovery enables me to
make the boiler absolutely tight and get any pressure I may
want. Only in this way it is possible to get some of the results
I published in 1900 (experiments I performed in 1899), and
only so can a plant be made to carry a voice, say, across the
Atlantic. As I wrote Mr. Morgan three years ago my present
plant will transmit speech over the Pacific with equal facility.

    I would like Mr. Morgan to feel that I am not
actuated by my own financial interest but only by the desire
to achieve for his sake. What little money is needed he might
perhaps advance simply as a small loan to the Company. Five
thousand dollars ($5000) would be ample provision for all

W.C.King--2

expenses during the next few months. Treasury shares might
be pledged until we get some returns. The shares cost at
present $175.00 in actual cash expended. So far I know
only a few of my own have been disposed of at that figure.

        Under inclosure I forward a letter, one of many
evidences on this nature, which please return.

                        Yours very truly,

JPM 0147 — transcript by Jeff Pearson

COPY.
Waldorf-Astoria
New York, Dec. 22, 1907

Mr. W. C. King,
23 Wall Street,
New York City, New York.

Dear Mr. King:

    Your letter relieves me, at least, on an harassing responsibility, but that is an unusual attitude to assume for a man of Mr. Morgan's broad views.

    In all I attempted for the past few years I have encountered a strange and firm resistance. Mr. Morgan will rather have the property go to the dogs than to help. I think I understand!

        Yours truly,

JPM 0148 — transcript by Jeff Pearson

June 9, 1908.

Messrs. J. P. Morgan & Company,

     23 Broadway.

         New York City.

Gentlemen:--

    Some crank has forwarded letters addressed to me to
your care, which have reached me apparently through your courtesy.

    You will greatly oblige me if you will henceforth throw
such letters in the waste-paper basket, where they belong.

Very truly yours,

JPM 0149 — transcript by Jeff Pearson

(COPY)

Waldorf-Astoria Hotel, March 31, 1913

J.P. Morgan & Co.,
      27 Wall Street
         New York, N. Y.

Dear Sirs:

     Please accept my heartfelt condolences on the
death of a great man who was the head of your famous firm.
When I can feel such a void in my heart and brain at the
passing of Morgan I can appreciate, in a measure, the depth
of feeling of those who were his lifelong comrades.

     The loss is irreparable but I hope that his wonder-
ful mind will further guide your house and that it will con-
tinue an example of banking on the highest plane, for the
honor of the country and the benefit of the whole world.

     Believe me,
               Very truly yours,
                    (Signed) N. Tesla

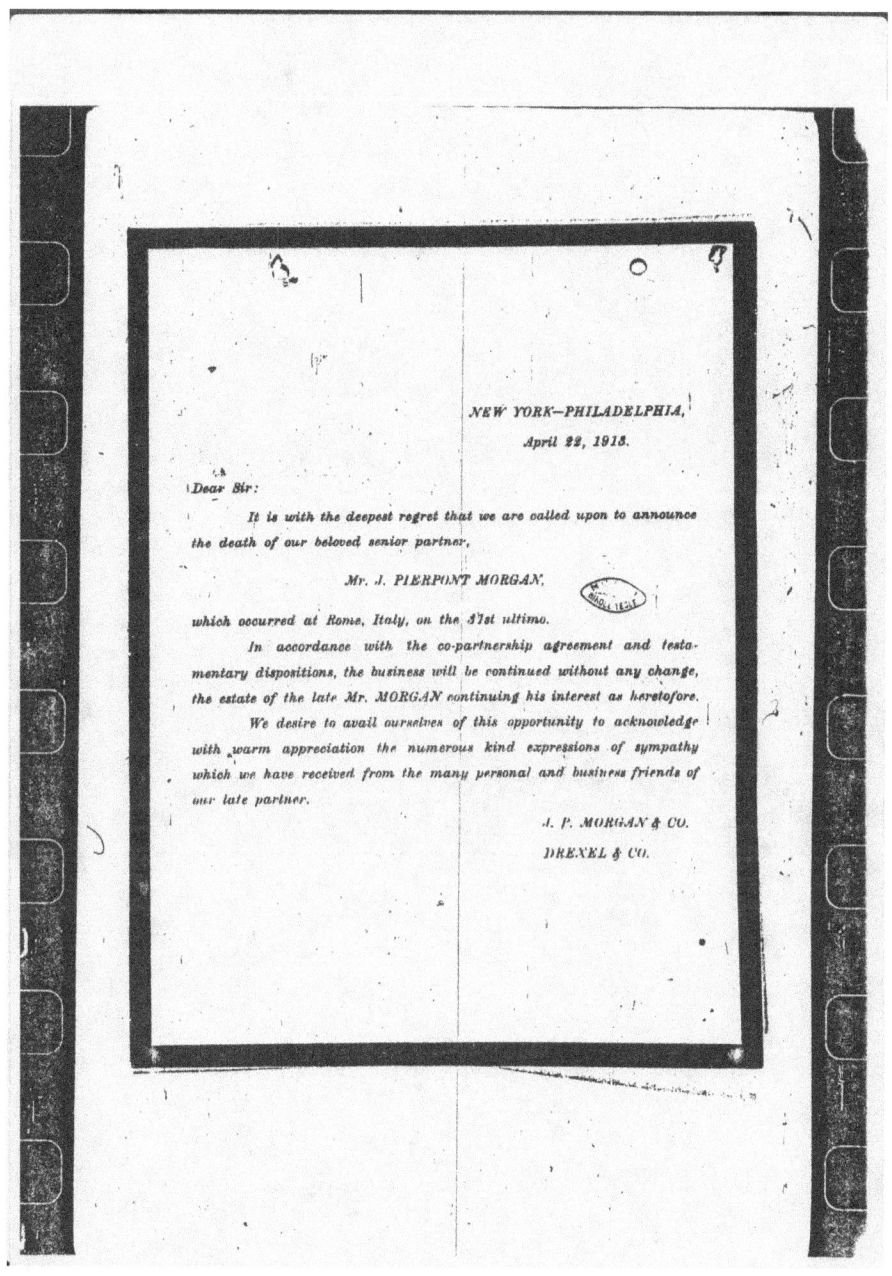

NEW YORK—PHILADELPHIA,

April 22, 1913.

Dear Sir:

It is with the deepest regret that we are called upon to announce the death of our beloved senior partner,

Mr. J. PIERPONT MORGAN,

which occurred at Rome, Italy, on the 31st ultimo.

In accordance with the co-partnership agreement and testamentary dispositions, the business will be continued without any change, the estate of the late Mr. MORGAN continuing his interest as heretofore.

We desire to avail ourselves of this opportunity to acknowledge with warm appreciation the numerous kind expressions of sympathy which we have received from the many personal and business friends of our late partner.

J. P. MORGAN & CO.

DREXEL & CO.

Tesla's public statements

Quote from "The Transmission of Electrical Energy Without Wires" by Nikola Tesla. Published in the "Electrical World and Engineer" on March 5th, 1904.
For a large part of the work which I have done so far I am indebted to the noble generosity of Mr. J. Pierpont Morgan, which was all the more welcome and stimulating, as it was extended at a time when those, who have since promised most, were the greatest of doubters. I have also to thank my friend, Stanford White, for much unselfish and valuable assistance. This work is now far advanced, and though the results may be tardy, they are sure to come.

In 1919 Tesla writes in his autobiography about this story:
I would add further, in view of various rumours which have reached me, that Mr. J. Pierpont Morgan did not interest himself with me in a business way but in the same large spirit in which he has assisted many other pioneers. He carried out his generous promise to the letter and it would have been most unreasonable to expect from him anything more. He had the highest regard for my attainments and gave me every evidence of his complete faith in my ability to ultimately achieve what I had set out to do. I am unwilling to accord to some smallminded and jealous individuals the satisfaction of having thwarted my efforts. These men are to me nothing more than microbes of a nasty disease. My project was retarded by laws of nature. The world was not prepared for it. It was too far ahead of time. But the same laws will prevail in the end and make it a triumphal success.

We now know better, yet Tesla did not want to discredit Morgan or speak bad about the dead.

June 5th, 1933, Letter by Nikola Tesla to the Editor of the New York Evening Post
Sir - Many of your readers, like myself, will feel indebted to you for your courageous and telling editorials relating to the investigation of the affairs of J. P. Morgan & Co. You have condemned these unfair proceedings in terms none too strong. Their undignified character is brought into evidence more and more, and it is becoming apparent even to the dullest observer that the honour and reputation of this famous banking house is resting on a foundation as solid as the Rock of Gibraltar. Perhaps it is fortunate that this investigation has been pushed so far, for in these times when confidence is most needed, the Morgans, in meeting these attacks, may be rendering the country service of inestimable value.
The general public has not even a remote idea of the position of this firm as a factor in the development of America. More than any other

force, they were instrumental in the furtherance of American interests throughout the world and in the building up of this country's power and prestige. Scores and scores of vast enterprises could not have been carried out but for their financial assistance. They helped Edison in commercializing his inventions and contributed to my own scientific researches with princely generosity. Edison and myself were only two among hundreds of inventors, engineers, artists and scientific men whose work they made possible. They advanced capital when all other doors were closed, stabilized the markets and fought depressions, not half-heartedly like others but with all their energies and resources, and at a peril to themselves. What they have added to national wealth staggers imagination.

I was intimately acquainted with the founder of this great house and know that his spirit is still with his successors. He set the example and they are endeavouring to emulate him with almost religious fervour. Persons worthy of respect can be found everywhere, but I have observed in the House of Morgan a largeness, nobility and firmness of character the like of which is very scarce indeed.

I can only smile when I read of the attempts to find something discreditable in the transactions of J. P. Morgan & Co. Not a hundred of such investigations will ever uncover anything which an unprejudiced judge would notconsider strictly honourable, fair, decent and in every way conforming to the high ideals and ethical standards ofbusiness. I would be willing to stake my life on it.

NIKOLA TESLA.

New York, June 2, 1933.

The End?

And this is where the battle ends. Tesla does contact J.P. Morgan jr. with some more business proposals but he does not ask for funds to complete Wardenclyffe. I think he felt it improper to ask his son to do something he so clearly opposed. Thus with the death of J.P. Morgan sr. the dream of Wardenclyffe also ends.

Or did it?

When Tesla returned from Colorado Springs he knew he had acquired world changing knowledge and he anticipated that he might not be able to complete his plan.
So he wrote an article for the Century Illustrated Magazine which was published in June 1900, titled "The Problem of Increasing Human Energy". In this article he used a code to convey his knowledge to a future experimenter.
As he writes:

Modern science says: The sun is the past, the earth is the present, the moon is the future. From an incandescent mass we have originated, and into a frozen mass we shall turn. Merciless is the law of nature, and rapidly and irresistibly we are drawn to our doom. Lord Kelvin, in his profound meditations, allows us only a short span of life, something like six million years, after which time the sun's bright light will have ceased to shine, and its life giving heat will have ebbed away, and our own earth will be a lump of ice, hurrying on through the eternal night. But do not let us despair. There will still be left upon it a glimmering spark of life, and there will be a chance to kindle a new fire on some distant star. This wonderful possibility seems, indeed, to exist, judging from Professor Dewar's beautiful experiments with liquid air, which show that germs of organic life are not destroyed by cold, no matter how intense; consequently they may be transmitted through the interstellar space. Meanwhile the cheering lights of science and art, ever increasing in intensity, illuminate our path, and marvels they disclose, and the enjoyments they offer, make us measurably forgetful of the gloomy future.

This article is his glimmering spark that will kindle a new fire in the distant future. And in March 1904 he wishes him (he who would find this code) success:

The following lines which, but for your initiative, might not have been given to the world for a long time yet, are an offering in the friendly spirit of old, and my best wishes for your future success accompany them.

With the purchase of this book you are supporting my research and increasing my hopes that one day the world will see a full-blown Magnifying Transmitter.

If you have not bought this book or you wish to support my project even more, please feel free to make a donation in one of the cryptocurrencies below.

BTC   LTC

Dash   Doge

THANK YOU!

www.ingramcontent.com/pod-product-compliance
Lightning Source LLC
Chambersburg PA
CBHW081601220526
45468CB00010B/2729